PRAISE FOR LUKE MARSH

A treasure trove of mind-blowing adaptations and surprising animal secrets—kid-friendly, fact-packed, and wildly engaging.

— DR. ELENA PARK, CHILDREN'S SCIENCE
EDUCATION ADVOCATE

This book makes the wild world approachable and thrilling; it's the perfect spark for independent readers and curious families.

— PROF. ORION REED, DEPARTMENT OF
COMPARATIVE ZOOLOGY, STELLAR
UNIVERSITY

From the deepest seas to jungle canopies, Incredible Animal Facts for Smart Kids turns learning into an adventure you can hold in your hands.

— DR. MAYA LIN, ZOOLOGIST AND PARENT
REVIEWER

INCREDIBLE ANIMAL FACTS
FOR SMART KIDS

INCREDIBLE ANIMAL FACTS FOR SMART KIDS

THE ULTIMATE COLLECTION OF FACTS ABOUT ANIMALS FOR CURIOUS MINDS

SMART KIDS WONDERS SERIES
BOOK ONE

LUKE MARSH

To every curious kid who asks how and why with a fearless heart, and to the grownups who feed that flame—teachers, parents, scientists—may this book spark wonder, fuel questions, and deepen compassion for all living beings. May Incredible Animal Facts for Smart Kids remind you that science is a celebration of curiosity, and that you, too, are capable of amazing discoveries.

The real voyage of discovery consists not in seeking new landscapes, but in having new eyes.

— MARCEL PROUST

CONTENTS

INTRODUCTION

Welcome to Incredible Animal Facts for Smart Kids, a wild tour through nature where every page hides something you never knew about animals.

This book is built for curious readers like you—fast, bite-sized facts, bright pictures, and plenty of 'wow' moments that make learning feel like an adventure.

From super-strong beetles to lightning-fast cheetahs, clever dolphins, and deep-sea oddities, you'll see how animals solve problems in surprising ways.

You'll meet record-breakers and master adapters—the fastest runners, the trickiest camouflagers, the loudest singers, and the brainiest problem-solvers.

Across themes like Super Senses, Animal Communication, Homes and Habitats, Nightlife, Ocean Wonders, Rainforest Wild, Desert Survivors, and more, you'll discover what makes each creature stand out.

Short reads with big ideas make it perfect for independent reading, classroom exploration, or sharing with family during car trips or bedtime.

Grab a page, ask a question, and let your curiosity roam— the animal world is full of surprises, and this book is your passport to notice, wonder, and conversation with friends.

ANIMAL RECORD BREAKERS

Welcome to Animal Record Breakers, where we meet the wild world's most extreme creatures. You'll discover the fastest sprinters, the biggest giants, the smallest wonders, and the cleverest adapters. Each fact shows how evolution turns clever ideas into amazing abilities.

1 - Cheetahs sprint up to 70 mph on short bursts.

2 - Peregrine falcons dive over 200 mph.

3 - Sailfish can swim at speeds around 68 mph.

4 - Giraffes reach about 16–18 feet tall.

5 - African elephants weigh up to 12,000 kilograms.

6 - Blue whales weigh up to 200 tons.

7 - Blue whale heart weighs as much as a small car.

8 - Bootlace worms can grow longer than 50 meters.

9 - Titan beetles reach lengths over 16 centimeters.

10 - Goliath beetles weigh up to about 100 grams.

11 - Bumblebee bats are the smallest mammals.

12 - Bee hummingbirds are about 5–6 centimeters long.

13 - Paedocypris is among the smallest vertebrates, under 8 millimeters.

14 - Paedophryne amauensis is about 7 millimeters long.

15 - Fairyflies are tiny wasps smaller than 1 millimeter.

16 - Dragonflies can fly about 35 miles per hour.

17 - Mantis shrimp have up to 16 color receptors.

18 - Box jellyfish venom can be deadly to humans.

19 - Inland taipans have venom among the most toxic of any land snake.

20 - Crocodiles have incredibly strong bite forces.

21 - Dung beetles can pull weights up to about 1,000 times their own body weight.

22 - Pronghorns reach about 55 mph for long distances.

23 - Arctic tern migrates about 25,000 miles round trip.

24 - Monarch butterflies migrate thousands of miles.

25 - Humpback whales are known for their elaborate songs.

26 - Octopuses can rapidly change color and texture.

27 - Giant squid have eyes as big as dinner plates.

28 - Blue whale tail span can exceed 25 feet.

29 - Saltwater crocodiles can reach lengths over 7 meters.

30 - Komodo dragons grow up to about 3 meters in length.

31 - Green anacondas can exceed 6 meters in length.

32 - Brown bears sprint up to around 40 mph.

33 - Peregrine falcons have an incredible hunting stoop.

34 - Blue whales produce very loud sounds.

35 - Platypuses and echidnas are monotremes.

36 - African elephants trunk contains about 40,000 muscles.

37 - Bombardier beetles spray boiling chemicals as a defense.

38 - Axolotls can regenerate lost limbs.

39 - Arctic foxes change their fur color with the seasons.

40 - Chameleons move their eyes independently.

41 - Poison dart frogs are tiny and highly toxic.

42 - Poison dart frog toxins are used by some peoples for hunting.

43 - Leopard seals are top predators in Antarctic waters.

44 - Electric eels can generate shocks around 600 volts.

45 - Electric organs run along the body of an electric eel.

46 - Saltwater crocodile ambush hunts are among the most stealthy predator tactics.

47 - Narwhals have long tusks that can grow over 2 meters.

48 - Male platypuses have venomous spurs on their hind legs.

49 - Sperm whales can dive for up to about 2 hours.

50 - Blue whale milk is extremely rich in fat, helping calves grow quickly.

51 - Ocean sunfish can weigh up to 2,000 pounds.

52 - Electric eels are actually knifefish, not true eels.

53 - Great white sharks sense electric fields with their ampullae of Lorenzini.

54 - Some snakes lay eggs, while others give birth to live young.

55 - Octopuses can fit through holes smaller than their bodies.

56 - Redback spiders have venomous bites that can cause intense pain.

57 - Poison frogs use bright colors to warn predators of their toxicity.

58 - Manta rays glide through the ocean with their broad fins.

59 - Baleen whales filter water using baleen plates to trap tiny prey.

60 - Orcas are intelligent toothed whales and top ocean predators.

61 - Albatrosses have very long wingspans among birds.

62 - Platypuses detect prey with electroreception in their bills.

63 - Chameleons camouflage and signal by color changes.

64 - Kangaroo rats extract water from seeds to stay hydrated.

65 - Octopuses have three hearts and blue blood.

66 - Camels store fat in their humps to endure long desert treks.

67 - Camels have long eyelashes and nostrils to protect from sand.

68 - Emperor penguins can dive to depths beyond 500 meters.

69 - Emerald tree boas blend with green foliage.

70 - Tiger sharks have stripes and sharp teeth.

71 - Ostriches run fast on two legs.

72 - Slow lorises have venomous bites.

73 - Porcupinefishes inflate and display sharp spines when threatened.

74 - Remoras hitch rides by attaching to larger animals.

75 - Sharks sense prey using electrical fields.

76 - Dolphins use echolocation to navigate and hunt.

77 - Pronghorns are built for endurance and speed.

78 - Many whale species migrate across oceans.

79 - Mantis shrimp strike with bullet-like speed and force.

80 - Peregrine falcons have superb eyesight to spot prey from above.

81 - The tuatara has a parietal eye.

82 - Octopuses can squeeze through tiny openings due to flexibility.

83 - Sea otters use rocks as tools to crack open shells.

84 - Toucans use their beaks to access fruit and signal health.

85 - Honeybees waggle dance to tell others where to find nectar.

86 - Some fish sense electric fields to locate prey in murky water.

87 - Blue whales feed on tiny krill.

88 - Electric rays stun prey with electricity.

89 - Narwhal tusks can sense environmental changes through nerves.

90 - Humpback whales breach and display their tails.

91 - Kangaroos hop with powerful hind legs to cover ground quickly.

92 - Crocodile eyes have a protective membrane that shields vision underwater.

93 - Cheetah tracks can identify distinctive individuals in the wild.

94 - Humpback whale songs evolve and spread across populations over time.

95 - Catfish use barbels to sense food in muddy water.

96 - A blue whale's tongue can weigh as much as an elephant.

2

SUPER SENSES: SEEING, SMELLING, AND HEARING THE UNSEEN

Animals don't experience the world the same way humans do—many have "super senses" that help them find food, avoid danger, and communicate. In this chapter, you'll discover creatures that can hear tiny sounds, smell invisible trails, see in near-darkness, and even sense electricity and heat.

97 - Owls locate prey by listening for faint sounds thanks to asymmetrical ears.

98 - Bats navigate and hunt by emitting ultrasonic calls and listening to echoes.

99 - Snakes have heat-sensing pits that detect warm prey in darkness.

100 - Mantis shrimps have eyes with many photoreceptors, giving exceptional color vision.

101 - Cuttlefish detect polarized light, aiding camouflage and signaling.

102 - Elephants rely on a strong sense of smell to find water, food, and friends over long distances.

103 - Bloodhounds track scent trails for long distances thanks to a large olfactory system.

104 - Dogs have far more olfactory receptors than humans, enabling detailed scent analysis.

105 - Star-nosed moles rely on their twenty-two nasal tentacles to explore textures by touch.

106 - Sea turtles navigate vast oceans by sensing the Earth's magnetic field.

107 - Electric eels sense prey by detecting electric fields in the water.

108 - Rays use electroreceptors to detect nearby electrical activity from hidden prey beneath sand.

109 - Whales and dolphins communicate with a rich repertoire of sounds heard underwater.

110 - Barn owls channel sound to their ears with facial discs for precise hearing.

111 - Pigeons can orient themselves and navigate using magnetic field cues.

112 - Dolphins can locate the source of sounds with precise auditory processing.

113 - Moles have highly sensitive whiskers that detect vibrations in soil.

114 - Turtles can see well in both water and air, aiding feeding and navigation.

115 - Beavers use tactile whiskers to sense water movement while foraging.

116 - Sharks can smell trace amounts of blood and other chemicals in water.

117 - Dogs rely on their powerful noses to identify people, objects, and scents.

118 - Wolves use scent marking and olfactory cues to communicate and navigate.

119 - Beagles, trained for scent tasks, demonstrate the extreme accuracy of canine smell.

120 - Bees detect flower scents and pheromones through their antennae.

121 - Butterflies see ultraviolet patterns on flowers that guide them to nectar.

122 - Dragonflies have eyes made of thousands of tiny lenses, giving them an almost 360-degree field of view and lightning-fast motion detection.

123 - Some birds can see ultraviolet light, helping them spot nectar guides on flowers and recognize kin by UV patterns.

124 - Seahorses rotate and move their eyes separately, watching for prey and predators at the same time.

125 - Reindeer can see ultraviolet light, which helps them spot food and predators in snowy landscapes.

126 - Goats have rectangular pupils that give them a wide, panoramic view to watch for danger while grazing.

127 - Owls' eyes are specially adapted for excellent night vision due to high rod density in the retina.

128 - Cats' eyes glow in the dark because of a reflective layer behind the retina called the tapetum lucidum.

129 - Some deep-sea fish can detect faint bioluminescence with highly sensitive retinas adapted to the darkness.

130 - Many animals can detect polarized light, which helps them navigate and detect prey or avoid glare.

131 - The eyes of seabirds are adapted to keep vision clear when diving into water and then resurfacing.

132 - Some fish see colors in blue-green water, helping them locate food and mates in their environment.

133 - The retina adapts to changing light by dilating or constricting the pupil, helping maintain good vision from shade to bright sun.

134 - Some birds can rapidly adjust focus to pick out a moving worm or insect on a branch.

135 - Octopuses have camera-like eyes capable of forming sharp images, aiding them in detecting shapes and textures.

136 - Insects with compound eyes can detect motion more quickly than humans, spotting predators far sooner.

137 - Some frogs can distinguish colors to locate ripe fruit or avoid dangerous plants.

138 - Many prey animals have eyes placed to give a wide field of view, helping them notice threats early.

139 - Some sharks can detect changes in light and shadow as a clue to finding prey in murkier waters.

140 - Certain birds have excellent color vision that helps them select ripe berries and avoid toxic ones.

141 - Some animals use color and brightness signals to communicate with others during courtship or warning displays.

142 - The ocular and brain systems work together to give creatures fast responses to visual clues in their environment.

143 - I hope you're having a great day!

3

ANIMAL ATHLETES: SPEED, STRENGTH, AND STAMINA

Get ready to cheer for the animal athletes of the wild. This chapter showcases speed, strength, and stamina across amazing creatures, from blazing sprinters to tireless travelers, with surprising, kid-friendly facts to spark curiosity.

144 - Lions can reach top speeds near 50 mph during chases.

145 - Leopards can run up to about 58 mph in quick bursts.

146 - Thomson's gazelles can reach 40–50 mph to outrun predators.

147 - Springboks can sprint around 40–45 mph across open plains.

148 - Greyhounds can reach around 40–45 mph on a track.

149 - African wild dogs can chase prey at around 40–45 mph in bursts.

150 - Red kangaroos can hop at roughly 40 mph in short bursts.

151 - Jackrabbits can bolt at speeds of 35–45 mph.

152 - Cheetahs balance at high speed with their long tails acting as rudders.

153 - Cheetahs have flexible spines that help extend their strides.

154 - The falcon's stoop uses gravity to boost speed and surprise prey.

155 - The Brazilian free-tailed bat can fly at speeds around 99 mph.

156 - Shortfin mako sharks can swim at speeds near 60 mph.

157 - Black marlin can burst through water at impressive speeds while chasing prey.

158 - Swordfish can accelerate quickly when pursuing prey through the water.

159 - Yellowfin tuna are known for fast bursts in open water.

160 - Dolphins can sprint in bursts of around 25 mph when chasing prey.

161 - Orcas can reach even higher speeds during hunting bursts.

162 - Penguins are powerful swimmers and can reach speeds around 6–12 mph.

163 - Albatrosses glide for hours, traveling thousands of miles with minimal effort.

164 - Humpback whales migrate thousands of miles between feeding and breeding grounds.

165 - Blue whales undertake long ocean journeys during migration.

166 - Salmon swim upstream against strong currents to reach their spawning grounds.

167 - Ants can carry loads many times heavier than their own body weight.

168 - Leafcutter ants move leaves many times heavier than themselves through teamwork.

169 - Rhinoceros beetles are incredibly strong for their size and can lift heavy objects.

170 - The rhinoceros beetle's horn helps it flip obstacles many times bigger than itself.

171 - Elephants can lift and move heavy objects using their trunks.

172 - The African elephant can sprint briefly and move with surprising speed when necessary.

173 - The American bison can run up to about 35 mph.

174 - Horses can reach speeds around 40–50 mph depending on breed and training.

175 - Zebras can run about 40 mph to escape predators.

176 - Gazelles can sprint at around 50 mph in open grasslands.

177 - The cheetah's powerful legs and flexible spine support rapid acceleration and speed.

178 - The leopard can accelerate quickly and maintain bursts to chase prey.

179 - The jaguar can sprint up to about 50 mph in short bursts.

180 - The tiger can sprint at around 40 mph for short distances.

181 - Wolves can run at speeds around 35–40 mph in bursts.

182 - Coyotes use fast bursts and agile maneuvers to evade threats and chase prey.

183 - Foxes use bursts of speed to escape danger and catch prey.

184 - Cougars can sprint with bursts of speed in mountainous terrain.

185 - The mountain lion's long legs help it cover ground quickly.

186 - The jaguar's stealth and power enable a sudden burst of speed to ambush prey.

187 - Snow leopards are masters of fast, agile chases across rocky slopes.

188 - The cheetah's big heart and lungs support extreme speeds.

189 - The cheetah's paw pads provide extra grip on smooth ground.

190 - The cheetah's tail helps stabilize turns at high speed.

191 - Swift birds are among the fastest in level flight and can cover great distances quickly.

192 - Albatross wings enable energy-efficient soaring over oceans for long journeys.

193 - Ospreys dive with speed and precision to catch fish.

194 - Kingfishers dart toward fish with rapid, direct flights.

195 - Hawks, falcons, and eagles rely on speed to overtake prey on the wing.

196 - Hummingbirds beat their wings incredibly fast, sometimes around 80 times per second.

197 - Archerfish knock prey off leaves by shooting precise jets of water.

198 - Octopuses can jet water to propel themselves quickly away from danger.

199 - Squid use jet propulsion to zip through water.

200 - Cuttlefish swim quickly and change color rapidly to confuse predators.

201 - Sea otters paddle with powerful, coordinated strokes for speed and maneuverability.

202 - Seals rely on strong flippers to propel themselves rapidly through water.

203 - Dolphins, porpoises, and other cetaceans can swim in tight, fast bursts when hunting or escaping.

204 - The crocodile can sprint on land in short bursts when threatened, surprising many observers.

205 - Alligators can move quickly across banks and shallow water when needed.

206 - Grizzly bears can sprint at surprising speeds for short distances.

207 - Brown bears can accelerate rapidly when threatened or pursuing prey.

208 - Wolves often hunt in packs, using endurance and speed to exhaust prey.

209 - Hyenas can run fast and with surprising stamina during hunts.

210 - Giraffes can run in bursts despite their height, using long legs to move quickly.

211 - Rhinoceroses can move with surprising speed when threatened, despite their bulky bodies.

212 - Hippos are surprisingly fast on land over short distances when angered.

213 - Elephants can move steadily and cover long distances in a day, showing great stamina.

214 - Camels can travel long distances in desert conditions with remarkable endurance.

215 - Llamas and alpacas can run briefly to escape danger or move quickly across terrain.

216 - Raccoons use quick bursts of speed to explore and scavenge safely.

217 - Mountain goats sprint along steep cliffs with agile leaps.

218 - Ibexes are adept at fast, nimble moves on rocky terrain.

219 - Caribou migrate long distances, demonstrating stamina across tundra.

220 - Elk can sprint quickly to dash away from threats.

221 - Moose can move rapidly through forests if threatened.

222 - Beavers swim quickly with strong, efficient strokes to evade predators.

223 - Otters are known for smooth, fast swimming while foraging.

224 - Seahorses swim slowly, but their strength helps they stay upright and steady.

225 - Bar-tailed godwits can fly over 7,000 miles non-stop in migration.

226 - Geckos' toe pads use microscopic hairs called setae to grip smooth surfaces, letting them sprint up glass and other slick surfaces.

227 - Cheetahs rely on semi-retractable claws that act like built-in cleats to grip the ground during their blistering accelerations.

228 - Hammerhead sharks have wide heads that stabilize their bodies and improve acceleration and turning speed when chasing prey.

229 - Kangaroos store elastic energy in the tendons of

their hind legs, making each bound more efficient and allowing long hops.

230 - Albatrosses endure long journeys through dynamic soaring, riding wind gradients to travel thousands of miles with minimal wing beats.

231 - Dolphins can rest one half of their brain at a time while swimming, helping them stay alert and keep pace during long chases.

232 - African wild dogs rely on extraordinary stamina and coordinated group chasing to exhaust prey over long distances.

233 - Orcas synchronize tail slaps and powerful tail strokes to accelerate quickly and corner fast-moving prey.

234 - Penguins use their flippers as hydrofoil-like limbs, generating strong underwater speed to pursue prey.

4

MASTERS OF DISGUISE:
CAMOUFLAGE AND MIMICRY

Creatures hide in plain sight using color, pattern, and shape. This chapter dives into camouflage and mimicry, from leaf-like insects to shape-shifting cephalopods, revealing how stealth helps animals hunt, avoid predators, and survive. Get ready to meet the masters of disguise in the wild.

235 - Leaf insects blend in by looking exactly like leaves, complete with vein patterns and jagged edges.

236 - Stick insects hide on branches by matching color, texture, and length to their surroundings.

237 - Leaf-tailed geckos have leaf-shaped tails that help them disappear on bark.

238 - The dead leaf butterfly rests with wings closed to resemble a dry leaf on a tree.

239 - Orchid mantises imitate real flowers, helping them lure pollinators as prey.

240 - Flower mantises use petal-like forelegs to resemble blossoms from which prey may approach.

241 - The twig snake can stay perfectly still among branches to remain unseen.

242 - Some snakes flatten their bodies to resemble leaves or sticks on the forest floor.

243 - Stonefish lie on the seabed and mimic the color and texture of rocks and coral.

244 - Flounders can rapidly change color and pattern to blend with the seabed as they move.

245 - Cuttlefish can change color, pattern, and even skin texture to match their surroundings.

246 - The mimic octopus can imitate several animals, such as sea snakes or crabs, to deter predators.

247 - Leafy sea dragons wear leaf-like fins to blend with seaweed and kelp forests.

248 - Pygmy seahorses cling to coral and take on the color of their coral backdrop.

249 - Nudibranchs often mirror the colors and textures of the corals and sponges they rest on.

250 - Some flatfish adjust hue and pattern to disappear against the ocean floor.

251 - Peppered moths illustrate camouflage through color morphs that match different tree backgrounds.

252 - Snow camouflage helps many animals stay hidden in wintery landscapes.

253 - Snowshoe hares switch fur color as the seasons change to stay concealed.

254 - Ptarmigans alter plumage color with seasons to blend into tundra habitats.

255 - Polar bears have translucent fur that appears white and helps them blend with ice.

256 - Chameleons can change color to blend with tree bark, leaves, or backgrounds.

257 - Some chameleons also change skin texture to resemble rough bark or lichen.

258 - Decorator crabs camouflage by attaching seaweed, shells, and other debris to their shells.

259 - Flower crab spiders match the color of the flowers they sit on to ambush pollinators.

260 - Praying mantises can resemble leaves or flowers to stay hidden while waiting to ambush prey.

261 - Some caterpillars mimic bird droppings to avoid being eaten on stems.

262 - Owl butterflies have large eye-like patterns on their wings to deter predators.

263 - Eyespots on wings can misdirect attacks toward wing margins rather than vital body parts.

264 - Glasswing butterflies have transparent wings that help them blend into flowers.

265 - Sea turtles can blend with seagrass and kelp beds when resting on the seabed.

266 - Some fish bury themselves in sand or mud to hide from predators.

267 - Mantis shrimps use camouflage to blend with coral backgrounds while hunting.

268 - Crabs and crustaceans often camouflage with shells and algae to hide from predators.

269 - Countershading, with darker tops and lighter undersides, helps animals hide from predators looking from above or below.

270 - Disruptive coloration uses bold patterns to break up an animal's outline so it's harder to spot.

271 - Masquerade camouflage makes an animal resemble an inedible object, like a leaf or twig.

272 - Batesian mimicry occurs when a harmless species imitates a harmful one to avoid predation.

273 - Müllerian mimicry happens when several harmful species resemble each other to reinforce warning signals.

274 - Cephalopods deploy chromatophores to rapidly change color and blend with backgrounds.

275 - Iridophores reflect light to create iridescent flashes that help break up outlines in the water.

276 - Leucophores contribute white patches that further muddle an animal's silhouette.

277 - Some animals use texture changes, via specialized skin or body structures, to imitate surfaces like bark or stone.

278 - Background matching is most effective on a uniform surface where colors are stable.

279 - Disruptive patterns often appear as stripes or spots that confuse a predator's sight lines.

280 - Masquerade disguises can cause predators to mistake an animal for an inedible object, delaying detection.

281 - Bioluminescent tricks in deep-sea species can aid camouflage by matching ambient light levels.

282 - Certain frogs and insects exhibit seasonal color shifts that align with changing environments.

283 - Some species employ micro-mleeve camouflage, where they blend with tiny background particles like sand or leaf debris.

284 - Chameleon skin changes can be influenced by mood, temperature, and social signals, not just background color.

285 - The texture of fur or scales can mimic the roughness of surfaces like bark, stone, or coral.

286 - Some crabs actively collect and attach materials from their surroundings to enhance their camouflage.

287 - Masquerade is not always perfect; some predators still detect camouflaged prey with patience and movement.

288 - Camouflage can be used by both predators and prey to increase success in hunting or evasion.

289 - Background matching is often most effective when the animal remains motionless for long periods.

290 - Disruptive coloration can be especially effective in open habitats with dappled light and shadow.

291 - Seasonal camouflage enables survival across changing climates and landscapes.

292 - In some ecosystems, camouflage also helps animals avoid overheating by presenting a cooler color blend.

293 - Mimicry can involve mimicking not just animals, but the textures and patterns of natural objects like leaves or rocks.

294 - Some species use camouflage to regulate their body temperature by matching sunlit or shaded backgrounds.

295 - The success of camouflage relies on the sensory perception of predators in the environment, including vision and motion detection.

296 - Certain camouflage forms can be learned by predators, making disguise an ongoing evolutionary arms race.

297 - Disclose: many camouflaged species rely on both color and movement to mislead predators.

298 - Mimicry sometimes requires the mimic to move in ways that resemble the model to ensure recognition by predators.

299 - Mulitple camouflage strategies can exist within the same species, depending on life stage or habitat.

300 - Some animals display seasonal camouflage by selecting pigments and patterns that fit current surroundings.

301 - Background matching often involves gradual changes in color that match the local environment.

302 - Disruptive coloration can be found in a wide range of animals, from insects to large mammals.

303 - Masquerade can involve patterns that resemble the texture of the background, such as leaf veins or bark grains.

304 - Camouflage can extend to sound and smell in some species, helping them avoid detection in murky environments.

305 - In the deep sea, light levels are so low that even small color changes can significantly improve concealment.

306 - Predators that rely on camouflage to approach prey often invest in motionless stalking rather than rapid chases.

307 - Some species use a combination of camouflaging and signaling to communicate while remaining unseen by others.

308 - The science of camouflage and mimicry spans biology, ecology, and behavior, revealing how animals survive across habitats.

309 - Color change in camouflage is orchestrated by specialized skin cells and nerve signals that control pigment distribution.

310 - Camouflage strategies can differ between day and night, adjusting to the visibility of each period.

311 - Disguise techniques are constantly evolving as

predators adapt and prey respond through natural selection.

312 - Mimicry of dangerous animals helps to deter predation by confusing or frightening potential predators.

313 - The study of camouflage helps scientists understand animal behavior, evolution, and the balance of ecosystems.

314 - For curious kids, camouflage shows how nature uses clever tricks to win at life without a single loud alarm.

315 - Learning about disguise helps us appreciate how diverse and creative life can be in even small corners of the world.

316 - From forest floors to coral reefs, disguise is a common and fascinating tool in the animal world.

317 - The more we learn about camouflage, the more we understand about how animals survive and thrive in challenging environments.

318 - Ant-mimicking spiders fool ants by matching both their body shape and their chemical scent, letting them pass through colonies unseen.

319 - Some octopuses raise skin papillae to create three-dimensional textures that resemble rough rocks or coral.

320 - In the deep sea, counter-illumination lets certain lanternfish and squid glow on their bellies to erase their shadows from below.

321 - Nightjars and other bark-dwelling birds use mottled plumage and subtle body tilts to blend in with weathered tree bark.

322 - Wobbegong sharks lie on the reef with irregular patchwork patterns that break up their outline and help them ambush prey.

323 - Some flatfishes adjust the size and placement of dark patches on their bodies so their skin texture mirrors the seabed as they glide.

324 - Cephalopods can rapidly coordinate color, brightness, and skin texture to blend with reef, sand, or seagrass in seconds.

325 - Cuttlefish can fine-tune micro-texture patterns on their skin to resemble the grain of sand or the roughness of rocks.

326 - Caterpillars and larvae sometimes imitate lichens or bird droppings on leaves to stay hidden from predators.

327 - The peppered moth shows seasonal camouflage by changing its wing color to match different trees across the year.

328 - Some fish use disruptive color patterns to break up

their shape and mislead predators about their true orientation.

329 - Aggressive mimicry is when a predator imitates harmless prey or a familiar animal to lure prey into a trap, a strategy observed in a few sharks and anglerfish.

5

BRAINS AND PROBLEM-SOLVERS

Meet nature's clever thinkers. This chapter dives into how animals use tools, remember details, plan ahead, and even play to solve problems, featuring dolphins, crows, elephants, and clever octopuses.

330 - Dolphins use sponges as tools to protect their snouts while foraging on the seafloor.

331 - Dolphins have signature whistles that function like names for individuals within a pod.

332 - Dolphins coordinate bubble-net feeding to herd fish into tight groups for easier catching.

333 - Dolphins can imitate new sounds and actions, helping them learn quickly from others.

334 - Dolphins remember friends and allies for many years, sometimes across generations.

335 - Dolphins learn hunting tricks by watching other dolphins and by teaching calves.

336 - Dolphins communicate with a rich array of clicks and whistles to share information about food or danger.

337 - Dolphins solve problems by thinking through steps to retrieve objects out of reach.

338 - Dolphins show cultural variation, with different communities using distinct hunting methods.

339 - Dolphins can understand human cues, such as pointing, when searching for hidden food.

340 - Dolphins display self-control in experiments that require waiting for a bigger reward.

341 - Dolphins remember where important foraging sites are located, sometimes across seasons.

342 - Dolphins exhibit social bonding through playful riding, synchronized swimming, and mutual grooming.

343 - Dolphins can plan ahead by choosing routes or tools based on past experiences.

344 - Dolphins can learn to imitate tricks demonstrated by humans or other dolphins.

345 - Dolphins display empathy, offering support to distressed pod members.

346 - Dolphins have brains that are highly folded, a feature linked to flexible thinking.

347 - Dolphins use tools to help with problem solving in some populations.

348 - Dolphins coordinate to help calves surface for air after dives.

349 - Dolphins display curiosity by exploring new objects and environments.

350 - Dolphins communicate through body language such as leaping, tail slaps, and fin shakes.

351 - Dolphins memory for social bonds helps them reunite with friends after long separations.

352 - Dolphins adapt their foraging strategies to different habitats, from rivers to open seas.

353 - Dolphins can pass learned behaviors from older to younger generations, showing cultural transmission.

354 - Crows bend wires into hooks to pull food from tight spaces.

355 - New Caledonian crows plan ahead by selecting the right tool for a future problem.

356 - Crows cache seeds in many hiding spots and remember each location for months.

357 - Ravens recognize individual humans and adjust their behavior based on past encounters.

358 - Corvids pass tool use tricks from elders to youngsters.

359 - Crows drop nuts on roads to crack them with passing cars.

360 - Ravens use deception to mislead rivals about hidden food sources.

361 - Some corvids stash extra tools near caches for later use.

362 - Crows solve multi-step puzzles that require using a tool, then a second tool.

363 - Ravens imitate unfamiliar sounds and invent calls to recruit mates or warn others.

364 - Elephants show self-awareness in mirror tests.

365 - Elephants remember water sources and migration routes across many years.

366 - Elephants use branches as tools to swat insects or reach high leaves.

367 - Elephants plan journeys to water and food, weighing distance and safety.

368 - Elephants coordinate to rescue a trapped herd member.

369 - Elephants learn new tricks by watching elders and copying success.

370 - Elephants remember kin and form long-lasting social bonds.

371 - Some elephants use tools to reach food beyond their reach.

372 - Octopuses open jars and bottle tops to reach food inside.

373 - Octopuses solve mazes in labs and remember the correct path after delays.

374 - Octopuses stash tools like coconut shells for later use.

375 - Octopuses learn by watching humans demonstrate tasks.

376 - In experiments, octopuses choose the correct tool from several options to win a reward.

377 - Octopuses explore unfamiliar objects with curiosity.

378 - Cuttlefish memorize past experiences and choose camouflage that worked before.

379 - Cuttlefish switch camouflage patterns quickly when backgrounds change.

380 - Parrots can solve cause-and-effect puzzles by testing actions and observing results.

381 - African grey parrots understand basic causality and predict outcomes in puzzles.

382 - Kea parrots are famous for curiosity and solving multi-step puzzles.

383 - Some parrots use tools to retrieve distant rewards after watching others.

384 - Parrots engage in play that helps practice problem-solving.

385 - Pigeons can learn to categorize objects by color and shape in lab tasks.

386 - Pigeons remember thousands of images and recall them after long delays.

387 - Dogs show cooperative problem solving when working with humans or other dogs.

388 - Dogs adapt their strategies when faced with new puzzles or obstacles.

389 - Wolves plan hunts and coordinate with teammates to catch prey.

390 - Wolves adjust tactics when prey changes direction or speed.

391 - Orcas coordinate complex group hunts and communicate with varied calls.

392 - Orcas remember successful hunting grounds and share knowledge with calves.

393 - Manatees show navigational memory to locate seagrass beds across seasons.

394 - Sea otters often wrap themselves in kelp to stay in place while feeding.

395 - Sea otters display social play that strengthens group foraging tactics.

396 - Spiders maximize prey capture by adjusting web design according to prey behavior.

397 - Spiders show problem-solving by reconfiguring webs when prey patterns change.

398 - Ants forage along dynamic trails and re-route when obstacles appear.

399 - Ants remember where food is located and return to those spots later.

400 - Ants practice tandem running, guiding a younger nestmate to a food source.

401 - Archerfish learn to shoot water to knock insects off plants and improve accuracy.

402 - Archerfish adjust their aim to compensate for water currents and moving prey.

403 - Chimpanzees use sticks to fish for termites and crack nuts, showing targeted tool use.

404 - Capuchin monkeys combine tools to retrieve rewards, demonstrating complex problem solving.

405 - Raccoons show flexible problem solving and can unlock simple latches to access food.

406 - Raccoons remember where they hid foods and avoid spots where rivals lurk.

407 - Crows can remember social networks and who helped whom in past foraging.

408 - Sea lions show problem solving by figuring out how to retrieve prey from nets.

409 - Beavers use dam-building as a long-term problem-solving strategy to manage water flow.

410 - Beavers remember the most successful dam-building sites across years.

411 - Dung beetles navigate using the Milky Way to roll dung balls in straight lines.

412 - Archerfish refine their water-shot technique with practice and adaptation.

413 - Owl monkeys illustrate complex social knowledge by remembering who shares resources.

414 - Hyenas coordinate with clan members to optimize hunting and feeding strategies.

415 - Squirrels cache nuts and remember many hiding spots through the winter.

416 - Octopuses can access multiple food sources by dexterously manipulating lids and openings.

417 - Spiders detect vibrations to decide where to strengthen networks and capture prey.

418 - Parrotlets and other small parrots show surprising problem-solving in toy experiments.

419 - Beehives adjust foraging strategies when flowers bloom at different times.

420 - Chimpanzees demonstrate planning by selecting tools that won't spoil before use.

421 - Octopuses can learn from mistakes quickly and modify strategies on the next trial.

422 - Crows minimize risk by sharing information about danger with flockmates.

423 - Ravens use social learning to adapt to changing food landscapes.

424 - Wolves communicate via scent and vocal cues to coordinate ambushes.

425 - Orcas can shift hunting groups for efficiency when prey type changes.

426 - The brain-to-body ratio of octopuses is unusually high among invertebrates, supporting advanced problem solving.

6

ANIMAL COMMUNICATION: CALLS, COLORS, AND CHEMICAL MESSAGES

Animals talk without words using sounds, colors, and scents to share information. This chapter explores the amazing ways creatures communicate, from songs and dances to scent trails and glowing signals.

427 - Birds sing long, intricate songs to defend territory and attract mates.

428 - Humpback whales produce songs that can travel for miles across the ocean.

429 - Elephants use low-frequency rumbles that carry long distances to stay in touch with distant family members.

430 - Prairie dogs give alarm calls that describe the type of predator and its distance.

431 - Fireflies flash species-specific light patterns to help individuals find mates.

432 - Cuttlefish rapidly change color, pattern, and texture to signal mood or intent.

433 - Peacocks display their large tail feathers to signal genetic quality to potential mates.

434 - Chameleons communicate mood and intent by shifting their skin color.

435 - Zebras use their stripes to recognize each other and coordinate group movement.

436 - Peacock spiders perform elaborate leg-waving dances to attract females.

437 - Ants lay down pheromone trails that guide nest-mates to food and back to the nest.

438 - Ants release alarm pheromones to warn others of danger.

439 - Bark beetles coordinate tree attacks using pheromones that recruit other beetles.

440 - Bottlenose dolphins use a mix of clicks and whistles to coordinate hunts and maintain group cohesion.

441 - Orcas have complex vocal repertoires that vary between pods.

442 - Some whale cultures keep songs that pass down through generations within a community.

443 - Bats use echolocation to navigate and hunt, and also use social calls to coordinate roosting and foraging.

444 - Electric fish generate electric fields that help them sense their surroundings and communicate with neighbors.

445 - Many fish change color rapidly to signal aggression, submission, or readiness to mate.

446 - Squid and cuttlefish use rapid color changes and patterns to signal intentions to rivals or mates.

447 - Fiddler crabs wave their oversized claw to advertise dominance and attract mates.

448 - Crickets and other insects produce sounds by stridulating, a process of rubbing body parts together.

449 - Frogs croak and trill in species-specific patterns that help individuals find partners.

450 - Toads emit calls during mating season to attract females.

451 - Sea urchins release pheromones to synchronize spawning events.

452 - Bioluminescent deep-sea creatures such as anglerfish use light signals to attract mates in the dark ocean.

453 - Some jellyfish use bioluminescence to signal potential mates in dim water.

454 - Fireflies use species-specific flash patterns to prevent cross-species mating.

455 - Cephalopods with light organs use glowing signals to communicate during courtship and schooling.

456 - Clownfish signaling within a group uses color patterns and body language to indicate social status.

457 - Parrots mimic sounds to communicate with flock members and learn social cues.

458 - Tigers communicate with a mix of vocalizations and body language including tail and posture signals.

459 - Gibbons sing long-distance duets that defend territories and strengthen family bonds.

460 - Birds perform dawn choruses to advertise territory and attract mates at sunrise.

461 - Bowerbirds build elaborate structures and decorate them with objects to signal creativity and fitness to females.

462 - Penguins rely on vocalizations to locate mates and chicks in colonies.

463 - Termites coordinate colony tasks with pheromones that guide workers to food and nest-building sites.

464 - Spiders communicate with vibrations through their webs to signal mates or warn intruders.

465 - Crabs and lobsters communicate through substrate vibrations and chemical cues to signal rivals and attract mates.

466 - Many birds can see ultraviolet light and use UV reflections in plumage as signals to potential mates.

467 - Dragonflies perform aerial courtship displays and body movements to attract mates.

468 - Salamanders use chemical cues in skin secretions to signal potential mates and rivals.

469 - Dolphins coordinate group hunts using whistle sequences, bubble streams, and clicking.

470 - Walruses use a broad repertoire of vocalizations and postures to communicate across ice.

471 - Bearded seals rely on loud calls to advertise territory and attract mates across distances.

472 - Sea lions use barks, grunts, and postures to coordinate foraging and maintain social order.

473 - Primates use facial expressions and vocalizations to convey emotions and social information.

474 - Meerkats use distinct alarm calls to signal different predators and guide group behavior.

475 - Lemurs rely on scent marking and calls to define group borders and attract mates.

476 - Orangutans emit long-distance calls to stay in touch with family and signal territory.

477 - Gorillas drum on chests and vocalize to communicate strength and group cohesion.

478 - Wolves howl to locate pack members and coordinate hunts.

479 - Hyenas use a range of vocalizations to share information about prey and social status.

480 - Bears show their fitness through behaviors and calls during courtship and territorial displays.

481 - Birds sing variations of songs to indicate experience and age to potential mates.

482 - Some lizards display bright throat patches during courtship to attract rivals and mates.

483 - Males of many species use calls and displays to defend territories against rivals.

484 - Freshwater fish may release pheromones after disturbance, signaling danger to others.

485 - Electric eels can send electric signals to communicate with other eels in the same water.

486 - Deep-sea lanternfish coordinate schooling using light signals that can be detected by others.

487 - Squid may adjust color patterns in synchrony to coordinate movement in a school.

488 - Seahorses communicate with mates through color changes and chemical cues in the water.

489 - Crows relay information about food caches through specialized calls and gestures.

490 - Parrots can learn to understand human cues and respond socially during captivity.

491 - Birds use colony-specific alarm calls to warn others of predators and coordinate movement.

492 - Cephalopods combine color signals with body patterns to share information with others.

493 - Chimpanzees use a toolkit of vocalizations, gestures, and facial expressions to coordinate group activity.

494 - Dolphins may use bubble signals and whistle patterns to coordinate play and foraging.

495 - Insects rely on chemical messaging at multiple scales, from individual recognition to colony-wide coordination.

496 - Ants can recognize nestmates by smell, helping them cooperate with their own colony.

497 - Red deer use loud roars to announce presence and attract mates.

498 - Tigers and other big cats leave scent messages through urine and scrapes to defend territory and signal rivals.

499 - Skunks use a strong odor as a warning signal to deter potential threats.

500 - Sea birds perform wing displays and tail flicks to signal dominance and readiness to breed.

501 - Some fish sense chemical cues to understand the social hierarchy in schools.

502 - Crabs detect vibration cues through the ground to locate mates or food.

503 - Electric rays can produce small electric pulses that serve as signals to other rays.

504 - Beetles release pheromones that help locate mates in the dark.

505 - Pufferfish create intricate circular sand patterns to attract mates and signal fitness.

506 - Jellyfish pulses and rhythmic movements can serve as cues about environmental conditions or prey presence.

507 - Dolphins adjust their whistle types when encountering new pod members or traveling long distances.

508 - Birds can learn new signals from other species when living in overlapping habitats.

509 - Bats share information about roost locations via social calls when migrating.

510 - Whale pods coordinate feeding using specific calls that indicate team-based strategies.

511 - Ant pheromones fade over time and need renewal to stay effective, so workers refresh them.

512 - Some animals use both color patterns and scent cues to ensure messages reach others in noisy environments.

513 - Fireflies can synchronize flashing within a group to boost mating success.

514 - Moths and other insects sometimes modify flight and scent emission patterns to avoid predators and attract mates.

515 - Frogs time their calls to environmental cues like rain and humidity to maximize mating success.

516 - Some species adjust their signaling strategies depending on social context to maximize success.

517 - In many species, tail flicks and body postures accompany calls to convey mood and intent.

518 - In crowded habitats, animals rely on a combination of colors, sounds, and scents to communicate clearly.

519 - The diversity of animal communication shows how evolution shapes senses to help life thrive.

520 - Some electric fish communicate with highly specialized patterns of electric organ discharges, using brief chirps and varying discharge shapes to signal aggression, courtship, or group coordination beyond simply detecting objects.

7

HOMES AND HABITATS: FROM BURROWS TO TREE CITIES

Dive into the wild world of animal architecture with this chapter, where living spaces are built from mud, moss, and mystery. From underwater dams to tree-top towns, animals design homes that protect, feed, and shelter their families while shaping the landscapes around them.

521 - Beavers build sturdy dams from sticks, mud, and stones to slow streams and create ponds that become bustling habitats.

522 - They also construct lodges with underwater entrances, giving beavers quick escape routes and winter safety.

523 - The ponds and wetlands created by beaver activity support frogs, fish, birds, and a wide array of insects.

524 - Termites construct towering mounds that house

millions of termites in climate-controlled nests beneath the soil.

525 - Some termite mounds reach heights taller than a person and include ventilation shafts that regulate temperature and humidity inside.

526 - Leafcutter ants cultivate fungus gardens inside elaborate underground nests, using leaf pieces as fertilizer for their food source.

527 - Ant nests are complex cities with chambers for brood, fungus gardens, food storage, and waste.

528 - Paper wasps and hornets craft papery nests by chewing wood fibers and mixing them with saliva, forming nurseries for offspring.

529 - These papery nests are often tucked under eaves, in tree hollows, or on the sides of buildings for safety.

530 - Weaver ants stitch leaves together with silk produced by their larvae to fashion living tree nests.

531 - The woven leaves form a sturdy, multi-layered home high in the canopy for a bustling ant colony.

532 - Orb-weaving spiders spin circular webs that serve as both homes and traps for prey.

533 - Some spiders decorate their webs with stabilimenta to camouflage the home or attract prey.

534 - Clownfish live among sea anemones, gaining protection from stinging tentacles while the anemone receives scraps and cleaning services.

535 - The anemone–clownfish relationship creates a mutual habitat on coral reefs.

536 - Coral reefs are built by tiny coral polyps that deposit calcium carbonate skeletons, constructing underwater cities for countless species.

537 - Coral polyps host symbiotic algae called zooxanthellae that share energy with the coral and color the reef.

538 - Reefs offer shelter for a dazzling variety of fish, invertebrates, and marine plants.

539 - Mangrove forests grow dense root systems that shelter juvenile fish and crustaceans while stabilizing shorelines.

540 - Mangrove roots slow water, trap sediment, and filter pollutants to create protected nurseries for life.

541 - Oyster reefs and mussel beds form hard underwater habitats that shelter juvenile fish and crustaceans.

542 - Mollusk shells left on beaches or in the sea can become homes for barnacles, hermit crabs, and other small organisms.

543 - Hermit crabs carry empty snail shells as portable homes and swap shells as they grow.

544 - Sea urchins hide in rocky crevices, contributing to the structure and diversity of tide pools.

545 - Octopuses use dens in caves, crevices, and abandoned objects to hide, molt, and hunt.

546 - Some octopuses rearrange rocks and shells around their den to improve camouflage and safety.

547 - Cuttlefish and other cephalopods dig shallow dens in sand to hide and ambush prey.

548 - Cephalopods can move objects around their den to alter camouflage and airflow.

549 - Coral reef walls host tiny creatures that live in cracks, crevices, and tiny caves within the reef architecture.

550 - Crabs and other crustaceans shelter in crevices and under rocks to dodge waves and predators.

551 - Tree hollows formed by decay or woodpeckers provide homes for owls, bats, and small birds.

552 - Woodpeckers carve cavities in dead trees, creating nest holes and roosts that other species reuse.

553 - Nuthatches, chickadees, and wrens often use natural cavities or nest boxes as year-round homes.

554 - Bluebirds frequently use pre-drilled nest boxes in gardens or woodlands as safe homes for their young.

555 - Wrens build compact nests in dense shrubs and hollow trees to shelter their families.

556 - Squirrels construct dreys, round leafy nests high in trees, as warm homes for their young.

557 - Weavers create large, communal nests in trees, sometimes forming tree cities with many families.

558 - Cliff swallows build mud nests on cliffs, bridges, and buildings for compact shelter for their colonies.

559 - Martins attach nests to ledges and eaves, creating bustling urban roosts for several generations.

560 - Kingfishers excavate vertical burrows in riverbanks to nest above the water and protect eggs.

561 - Herons and egrets build tall stick nests in trees in wetlands, often sharing space with others in colonies.

562 - Gannets and boobies nest on cliff tops or remote islands, forming large, noisy colonies.

563 - Puffins dig burrows on cliff faces to cradle their eggs away from gulls and wind.

564 - Puffin nests are lined with grass and moss to keep eggs warm and safe.

565 - Swiftlets build nests entirely from edible saliva inside caves, forming sturdy roosts for thousands of birds.

566 - Humans harvest swiftlet nests for cuisine, impacting wild populations and habitats.

567 - Bats roost in caves, hollow trees, and even man-made structures, gathering in maternity colonies.

568 - Bat roosts regulate temperature and protect roosting bats from predators.

569 - Fruit bats rest in tree canopies and caves, driving seed dispersal and forest regeneration.

570 - Some bats roost in urban buildings, turning cities into important roosting habitats.

571 - Naked mole-rats live in vast underground colonies with specialized rooms for breeding, nurseries, and stores.

572 - Their subterranean city stays humid and cool, perfect for long-term life underground.

573 - Gopher tortoises dig long burrows that shelter themselves and many other species when abandoned by the tortoise.

574 - Meerkats live in cooperative burrow networks with sentinels guarding the entrance.

575 - Meerkat burrows include nursery rooms and lookout posts at the entrance for safety.

576 - Badgers dig deep setts with multiple entrances that stretch underground for protection and rest.

577 - Badger setts can become homes for other species when abandoned, becoming mini ecosystems.

578 - Aardvarks dig burrows that break through tough ground, providing shade and shelter for themselves and other animals.

579 - Wombats dig sturdy burrows with many rooms through their grassy habitats for safety and sleep.

580 - Desert foxes such as the fennec dig cool, shaded burrows to escape daytime heat and conserve water.

581 - Dingoes and other desert mammals also use burrows to hide and raise young when possible.

582 - Polar bears use snow dens and hollow ice to shelter cubs during storms and harsh weather.

583 - Arctic foxes dig snow burrows that trap warm air and protect pups in freezing winters.

584 - Hares and other small mammals use burrows under snow as escape routes from predators.

585 - Bats that roost in caves contribute to a unique nocturnal ecosystem that thrives in darkness.

586 - A colony of bats in a cave often hosts dozens to millions of individuals depending on species and cave size.

587 - Birds that nest in tree cavities rely on older, decaying trees or natural hollows as homes for breeding.

588 - Nesting birds often reuse old nests or abandon nests that become home to insects and scavengers.

589 - Bluebirds, chickadees, and nuthatches frequently adapt nest boxes in gardens to create reliable homes.

590 - Owls commonly use tree cavities, abandoned nests, or nest boxes placed by people as safe nesting sites.

591 - Kookaburras and other woodland birds nest in hollow trees, using the natural architecture of their habitat.

592 - Hummingbirds build tiny cup-shaped nests tucked into shrubbery with spider silk to hold them in place.

593 - Parrots in tropical forests carve out tree cavities or reuse old holes for breeding.

594 - Parrot nests can be located high in trees to avoid ground predators.

595 - Ravens and crows sometimes use cavities or urban structures as homes for their young.

596 - Nests in trees provide shade, warmth, and protection from many predators for generations of birds.

597 - Tree cavities are also valuable winter refuges for small mammals and invertebrates when conditions tighten.

598 - Tree cities—clusters of nests high in one tree or a grove—offer a remarkable example of animal urbanism.

599 - Weaver birds create dense nest clusters that look like living neighborhoods in the canopy.

600 - Cliff-dwelling birds construct nests on exposed ledges to minimize predator risk while maximizing wind exposure for egg care.

601 - Nesting colonies on cliffs can be sites of intense vocal activity and healthy genetic mixing due to close proximity.

602 - Nesting plateaus on large seabird colonies provide stable access to food sources and sheltered spaces for chicks.

603 - Some shorebirds nest in shallow scrapes on sandy beaches that blend into the shoreline for protection.

604 - Bridge and building ledges have become modern homes for many urban birds, turning cities into new habitats.

605 - Nests lined with feathers, grasses, and plant fibers help keep eggs warm and dry during cool nights.

606 - Animals often share habitats, with burrow networks or hollow trees hosting many different species over time.

607 - Tree hollows and nest boxes are essential for maintaining bird diversity in forests and suburbs.

608 - Habitats built by animals often influence plant

communities by altering light, moisture, and soil structure.

609 - Urban trees with nesting cavities can become vital green infrastructure for birds and insects.

610 - The sustainability of these habitats depends on preserving old trees, fallen logs, and natural cavities in landscapes.

611 - Studying animal habitats reveals how life adapts to every niche, from deserts to oceans.

612 - Seabird colonies enrich island soils with guano, turning bare cliffs into lush microhabitats for plants, insects, and scavengers.

613 - In tropical forests, bromeliads hold rainwater in their leaf cups, creating dozens of tiny ponds that frogs use to lay eggs and raise tadpoles.

614 - The size, depth, and placement of a tree hollow determine which animals can use it, turning each hollow into its own little ecosystem.

615 - Coral reefs are like underwater apartment complexes, with branches, holes, and overhangs that host crabs, shrimps, and countless small fish.

616 - Burrowing mammals aerate soil and mix in seeds, creating above-ground patches of new plant life and homes for insects and rodents.

617 - Wetland edges formed by reeds and grasses slow water, trap sediment, and create sheltered nurseries where amphibians, insects, and small fish can grow.

618 - Abandoned bird nests and insect nests often become mini homes for spiders, beetles, mosses, and other life, keeping habitat diversity alive year after year.

8

NIGHTLIFE: CREATURES
OF THE DARK

Night falls and a different world wakes up. In this chapter you'll meet nocturnal animals and learn how they see in the dark, hear hidden sounds, and hunt with amazing stealth.

619 - Owls can rotate their heads about 270 degrees to scan the night without moving their bodies.

620 - Bats navigate and catch prey in total darkness by using echolocation.

621 - Fireflies glow at night with bioluminescence to attract mates.

622 - The fennec fox's giant ears help it hear tiny prey and keep cool in the desert heat.

623 - Some moths have ears that detect the calls of bats, helping them escape danger.

624 - Nocturnal eyes contain many rod cells that see better in low light.

625 - The aye-aye taps wood with its long finger to locate tasty grubs inside.

626 - Nightjars are night-flying birds that catch insects in flight.

627 - Anglerfish lure prey with a glowing lure in the dark ocean depths.

628 - Vampire bats sense heat to locate blood vessels on sleeping animals.

629 - Domestic cats are often most active at night and around twilight.

630 - Geckos have adhesive toe pads that let them climb walls in the dark.

631 - Raccoons use their sensitive whiskers and nimble paws to explore in the dark.

632 - A tapetum lucidum behind the retina helps many nocturnal animals see in dim light.

633 - Owls' soft wing feathers help them fly silently.

634 - Nocturnal animals commonly rely on smell and hearing when vision is limited.

635 - The deep sea glows with bioluminescence, lighting up the black water.

636 - Some deep-sea fish use glow to attract prey in the deep water.

637 - Nectar-feeding bats help pollinate night-blooming flowers.

638 - The eyes of many nocturnal mammals are large to collect more light.

639 - Many nocturnal animals rest in burrows, caves, or tree hollows during the day.

640 - Crickets and other night insects use sound to attract mates in the dark.

641 - The moon and stars can help some animals navigate during night travel.

642 - Nighttime helps some predators avoid daytime heat and crowds.

643 - Some insects sense vibrations in the air to locate approaching predators.

644 - Bats can eat hundreds of insects in a single night, helping farmers reduce pests.

645 - Some owls can locate prey by hearing the slightest movement under snow or leaves.

646 - Nighttime is when many desert mammals come out to hunt and drink.

647 - A nocturnal animal might have a bigger pupil to gather more light.

648 - The echolocation calls of bats are ultrasonic, above human hearing.

649 - Some fish and squid provide light-producing organs to communicate in dark water.

650 - Many nocturnal predators hunt by ambush, using cover to stay hidden.

651 - Some animals can run surprisingly fast at night despite low light.

652 - Bioluminescent bacteria sometimes power the glow of deep-sea life.

653 - Some owls hunt by listening for the tiniest rustle of prey moving on the ground.

654 - Nocturnal animals often have flexible sleep schedules to fit resources.

655 - The whiskers of many mammals help them feel the environment in the dark.

656 - Some flowers open at night to attract night-pollinating creatures.

657 - The deep-sea amphipods glow to communicate with others.

658 - Nighttime weather changes can alter how easily animals hear and see.

659 - Some animals use echolocation in both air and water.

660 - The eyes of cats reflect light at night, creating a glowing effect in flashlights.

661 - In the dark, many animals rely on smell and touch more than sight.

662 - The big eyes of tarsiers are a hallmark of their night activity, helping them see in dim light.

663 - Some nocturnal insects use pheromones to coordinate mate finding.

664 - In the Arctic, some animals time their activity around the long polar night to feed.

665 - The eye-shine seen when light hits eyes at night comes from the tapetum.

666 - Owl pellets show what they have eaten and can reveal prey.

667 - Some nocturnal animals migrate to follow food sources across landscapes.

668 - Nighttime temperatures influence which animals come outside to hunt or forage.

669 - Some nocturnal birds can migrate long distances at night using stars as guides.

670 - Some animals communicate by scent signals at night to coordinate movement.

671 - The ears of many nocturnal mammals are large relative to their head.

672 - Some bats use their wing membranes to help maneuver and reduce noise.

673 - The retinas of nocturnal animals are often rich in rods.

674 - Some sea creatures glow to lure prey.

675 - Nighttime insect populations can be abundant in warm months.

676 - The nocturnal forest is full of chittering and rustling.

677 - The owl's ability to locate prey by sound reduces reliance on bright light.

678 - Bats often roost in caves, trees, and even buildings during the day.

679 - Some nocturnal animals travel through underground tunnels to stay hidden.

680 - The eyes of nocturnal birds are often large for their body.

681 - Nocturnal animals often tolerate cooler night temperatures better than daytime animals.

682 - The sense of smell helps nocturnal animals track prey after rain.

683 - Some animals can see ultraviolet light, which can help them see markings even in dim light.

684 - Certain geckos can regrow lost tails after a night-time predator attack.

685 - The tides influence the behavior of many nocturnal marine animals.

686 - Some nocturnal animals use scent trails to guide others to food.

687 - In the dark, many animals rely on feel and touch more than sight.

688 - Some nocturnal birds produce calls that travel long distances at night.

689 - The night is when many migrant birds move to warmer climates.

690 - Nocturnal mammals often have rapid, precise movements to stay hidden.

691 - Fireflies flash in different colors and patterns to communicate.

692 - The deep sea uses glow to communicate across vast darkness.

693 - Some nocturnal fish use lateral lines to sense vibrations in water.

694 - The ears of nocturnal predators help them locate prey in the dark.

695 - Some animals use decoys to mislead predators at night.

696 - Nocturnal insects are abundant in tropical forests at night.

697 - The night is a prime time for many frogs and toads to call for mates.

698 - Some desert animals obtain water from the moisture in their food rather than drinking.

699 - Some penguins in polar regions remain active at night during the polar night.

700 - Moonlight can help certain animals with balance and navigation.

701 - The ears of nocturnal predators are often located on the sides of the head to detect sounds from different directions.

702 - The night is a stage for many amphibians to call and mate in ponds.

703 - Bioluminescent algae can light up waves at night.

704 - Nocturnal animals often have stronger senses of hearing than day-active animals.

705 - Some birds of prey hunt at night by using their eyes to see in dim light.

706 - Bioluminescence is produced by chemical reactions inside specialized cells.

707 - Even in darkness, sound travels differently across air, water, and ground.

9

OCEAN WONDERS: GIANTS, SPRINTERS, AND DEEP-SEA STRANGERS

Dive into the wild world of ocean wonders where giants swim, swift sprinters flash through the blue, and deep-sea strangers glow in the dark. This chapter explores how marine life survives crushing pressure and near-freezing darkness, from colossal giants to glow-in-the-dark hunters and strange deep-sea residents.

708 - The blue whale is the largest animal ever known to have lived on Earth.

709 - Blue whales can eat up to four tons of krill in a single day during peak feeding.

710 - They reach lengths of up to 100 feet, making them the giants of the ocean.

711 - Blue whales migrate thousands of miles between polar feeding grounds and warmer breeding waters.

712 - Whale songs can travel long distances underwater and may be heard by other whales far away.

713 - Blue whale skin is thick and dark, helping it blend with deep water when seen from above.

714 - The fin whale is the second-largest whale and can reach around 70 feet in length.

715 - Whale sharks are the largest fish in the world, not whales.

716 - Whale sharks can grow to about 40 feet long and weigh several tons.

717 - They swim with mouths wide open, filtering water through their gills to capture plankton.

718 - Whale sharks have distinctive white spots and unique patterns that scientists use to identify individuals.

719 - Giant squid can reach lengths of about 13 meters from tip of tentacles to mantle.

720 - Giant squid have the largest eyes in the animal kingdom.

721 - Giant squid arms and tentacles are lined with suction cups and sometimes curved hooks.

722 - Colossal squid can weigh up to about 500 kilograms and live in cold Antarctic waters.

723 - Colossal squid have two long feeding arms in addition to their eight arms.

724 - The beak of a colossal squid is extremely strong and can pierce prey.

725 - Giant and colossal squids were once shrouded in legend before scientists studied them.

726 - Baleen whales filter food using comb-like baleen plates made of keratin.

727 - The vast mouths of baleen whales can take in enormous amounts of water when feeding.

728 - Whales are built for long migrations rather than quick sprints.

729 - The whale shark feeds by filtering plankton-rich water through its filtering system.

730 - Whale sharks have broad heads and tiny teeth specialized for filtering plankton.

731 - Whale sharks' skin features a pattern of white spots that looks like a fingerprint.

732 - The giant squid is mostly elusive and rarely seen at sea.

733 - Colossal squid have a beak and hooks on their arms to grab prey.

734 - Blue whales store energy as blubber, which helps keep them warm in cold seas.

735 - Humpback whales are famous for long migrations, complex songs, and acrobatic breaches.

736 - Orcas, also called killer whales, are large dolphins that can swim very fast.

737 - Orca pods work together to herd fish and even create waves to dislodge prey.

738 - Blue whales have incredibly slow metabolisms, which helps them thrive on tiny prey.

739 - Narwhals are sometimes called the unicorns of the sea because of their long tusk.

740 - A narwhal's tusk is an elongated tooth used in social and environmental sensing.

741 - The giant Pacific octopus can grow to be very large, sometimes over 16 feet across.

742 - Octopuses are masters of camouflage and can squeeze through tiny gaps.

743 - Octopuses can eject ink to confuse predators while escaping.

744 - The giant squid (Architeuthis) has a complex nervous system and large eyes.

745 - In many anglerfish species, males fuse with females and become essentially parasitic mates.

746 - Lanternfish are small deep-sea fish that glow due to bioluminescence.

747 - Bioluminescence in the deep sea helps creatures attract prey and signal others.

748 - Flashlight fish have glowing organs that help them see and avoid predators.

749 - Some deep-sea jellyfish glow softly to attract prey or mates.

750 - Fangtooth fish have very large teeth relative to their body size.

751 - Dragonfish glow with photophores to camouflage and lure prey.

752 - Deep-sea life relies on energy-efficient bodies and slow metabolisms to cope with scarce food.

753 - Goblin sharks can protrude their jaws to snatch prey in the dark.

754 - Many deep-sea creatures have soft, gelatinous bodies to withstand high pressure.

755 - Pressure at the deepest depths makes most shallow-dwelling animals unable to survive.

756 - Yeti crabs carry bacteria on their hairy claws that harvest nutrients in cold waters.

757 - Dumbo octopuses get their name from ear-like fins that flap as they swim.

758 - The vampire squid can turn on its bioluminescent lights to escape predators.

759 - Frilled sharks resemble ancient, primitive sharks and can grow large.

760 - Viperfish have long fangs and hinged jaws for catching prey in darkness.

761 - Blobfish have a gelatinous body that helps them survive in high pressure.

762 - Snailfish have soft skeletons that stay flexible in deep water.

763 - Some cusk-eels can live at extreme depths with slow metabolisms.

764 - Gulper eels have enormous mouths that can swallow prey larger than themselves.

765 - Many deep-sea creatures rely on sensing vibrations and chemical cues to find prey in the dark.

766 - Copepods are tiny crustaceans that form a huge part of the deep-sea food web.

767 - Some jellyfish have long stinging tentacles that reach far from their bodies.

768 - Reproductive strategies in the deep sea often involve low-energy, energy-efficient patterns.

769 - Bioluminescent bacteria can glow on or inside animals, creating light from within.

770 - Lanternfish produce light along their bodies and tails for camouflage.

771 - Anglerfish live in deep pelagic zones far from light and rely on their lure to attract prey.

772 - In the deep sea, sunlight fades within a few hundred meters, creating a world of permanent darkness.

773 - Octopus arms have suction cups that help anchor and grab objects.

774 - The giant squid's eyes are among the largest in the animal kingdom and help it detect faint light.

775 - The barreleye fish has a transparent head that allows it to look upward while its eyes face forward.

776 - Glass squid have transparent bodies, making parts of them nearly invisible in certain light.

777 - Fangtooth fish are dark-colored, blending into the deep-black environment.

778 - The vampire squid uses a cloak-like web and fins to move and to escape predators.

779 - Dragonfish have sharp teeth and rely on bioluminescence for hunting.

780 - The deep sea supports a surprising diversity of life, including crabs, shrimp, and octopuses.

781 - The deepest ocean depths experience about 1,000 times more pressure than the surface.

782 - Bioluminescence is common in the deep sea and is used for many purposes.

783 - Many deep-sea animals tolerate near-freezing temperatures.

784 - Hydrothermal vents host communities powered by chemosynthesis rather than photosynthesis.

785 - Bacteria near vents form the base of the food web for vent-dwelling animals.

786 - Giant tube worms can grow to over 2 meters and host symbiotic bacteria.

787 - Vent ecosystems rely on chemical energy from minerals rather than sunlight.

788 - Octopuses have beaks that can crack tough prey like crabs.

789 - Some octopuses are famous for clever problem-solving tasks in experiments.

790 - The blue-ringed octopus carries venom that can be deadly to humans.

791 - The deep sea often relies on detritus falling from above to supply energy for life.

792 - Photophores are light-producing organs found on many deep-sea fish.

793 - Sperm whales dive deep to feed on squid and can stay underwater for long periods.

794 - Giant isopods are among the ocean's largest crustaceans and can grow to over a foot long.

795 - Blue whales communicate with low-frequency sounds that can travel across oceans.

796 - Deep-sea vents host thriving communities where life thrives under extreme conditions.

10

RAINFOREST WILD: JUNGLE GIANTS AND TINY TREASURES

Rainforest Wild: Jungle Giants and Tiny Treasures dives into the layered world of tropical forests where giants and tiny creatures share every leaf and twig. From towering trees to tiny leaf insects, every level hides wonders waiting to be discovered.

797 - The rainforest has four layers: emergent, canopy, understory, and forest floor.

798 - Emergent trees rise above the rest of the forest to catch the sun.

799 - The canopy forms a leafy shield that shields the lower layers from direct light.

800 - The understory stays warm and humid, hosting many secretive animals.

801 - The forest floor is dark and damp, where fallen leaves feed the soil.

802 - Rainforests cover a small portion of Earth yet host more than half of its plant and animal species.

803 - A single hectare can hold dozens of tree species and thousands of insects.

804 - Leafcutter ants cut leaves to farm fungus that feeds the whole colony.

805 - Some poison dart frogs secrete potent toxins that deter many animals.

806 - Jaguars are powerful hunters that stalk prey in the dense undergrowth and along riverbanks.

807 - Jaguars are skilled swimmers and will prey on fish, caimans, and even capybaras when they can.

808 - Harpy eagles have enormous talons and can lift prey larger than many other birds can handle.

809 - Macaws are vividly colored parrots that travel in noisy, social flocks through the canopy.

810 - Toucans use their oversized bills to reach fruit and to help regulate body heat.

811 - Sloths move slowly to conserve energy in the warm, humid rainforest environment.

812 - Sloth fur hosts algae, giving them a greenish tint that helps camouflage them on trees.

813 - Tree frogs often have toe pads that grip leaves and bark and help them climb.

814 - Poison dart frogs lay eggs in damp places and carry tadpoles to water-filled tree holes or pools.

815 - The Amazon basin harbors millions of insect species, many of them unseen by most people.

816 - The jaguar's bite is one of the strongest among mammals, capable of cracking turtle shells and skulls.

817 - Green anacondas are among the longest snakes and hunt largely by ambush in rivers and swamps.

818 - Piranhas swim in swift, tightly packed schools that can strip prey quickly under the right conditions.

819 - Bullet ants produce a sting so painful that people describe it as similar to being shot, hence the name.

820 - The blue morpho butterfly appears electric blue when rays of sunlight strike its wings.

821 - Poison darts frogs obtain their toxins from the insects they eat in the wild.

822 - Capybaras, the largest rodents, live in social groups near waterways and take long naps in the shade.

823 - Capybaras use soft vocalizations and scent cues to stay in touch with family members.

824 - Orangutans swing through Southeast Asian rainforests using long arms and powerful grasps.

825 - Orangutans build sleeping nests in trees each night to rest safely above the forest floor.

826 - Plant-animal partnerships are everywhere, from pollinators to seed dispersers that shape the forest.

827 - Epiphytes such as bromeliads and orchids perch high on tree trunks, collecting rainwater and debris.

828 - Bromeliads create mini-water ecosystems that frogs and insects use as temporary pools.

829 - Fig trees rely on tiny wasps for pollination, linking figs and wasps in a mutual relationship.

830 - Orchids often depend on specific pollinators, sometimes tricking them with cunning flower shapes and scents.

831 - The rainforest canopy hosts colorful birds like macaws, toucans, and trogons that rely on fruit.

832 - Bats help pollinate flowers and disperse seeds, keeping the forest alive after dark.

833 - Tapirs wander forest trails, eating fruit, leaves, and shoots, and helping disperse seeds through their droppings.

834 - Giant otters patrol rainforest rivers, forming family groups that hunt together.

835 - The Amazon River and its tributaries are home to countless species, including many fish, reptiles, and birds.

836 - Caimans lie in wait along river edges, ambushing fish or young animals that come to drink.

837 - Caiman skin patterns camouflage them among riverbank shadows and reflections.

838 - Frog skins can be slippery and require careful handling if touched, especially poison darts.

839 - Tree frogs skillfully climb using suction-cup toes that grip smooth surfaces.

840 - The hummingbird is a small, fast-flying nectar feeder common in many rainforests.

841 - Nighttime in the rainforest brings a chorus of frogs, crickets, and owls that fill the air.

842 - The agouti is a forest rodent that forages on the ground and helps scatter seeds.

843 - Orchids and other epiphytes provide microhabitats for insects, birds, and frogs high in the trees.

844 - The rainforest stores large amounts of carbon in its trees and soil, helping regulate the climate.

845 - The warm, wet climate of tropical rainforests supports fast plant growth and year-round fruit.

846 - Bullet ants are known for their intense sting and for traveling in long, organized trails.

847 - Army ants form raiding columns that sweep through the forest floor and feed on other insects and small animals.

848 - Some rainforest insects can see in the ultraviolet spectrum, revealing patterns invisible to humans.

849 - Glass frogs have transparent bellies, allowing scientists to study their internal organs.

850 - The rainforest hosts many species of primates, including howler monkeys and capuchin monkeys.

851 - Howler monkeys make loud calls that travel through the trees and help keep bands together.

852 - The rainforest supports countless species of snakes, including boas and pit vipers, that blend into their surroundings.

853 - The rainforest roof often shadows streams and pools, keeping water temperatures stable for aquatic life.

854 - Some tree frogs can change color to blend with their surroundings.

855 - The near-equatorial rainforest experiences little seasonal change, so life stays busy year-round.

856 - Many rainforest animals rely on fruit trees for a reliable food source, making fruiting cycles critical.

857 - Seed dispersal by animals helps forests regenerate after disturbances like storms or tree falls.

858 - The fig wasp is a tiny insect that must complete its life cycle inside the fig fruit.

859 - The canopy hosts many parrots, toucans, and nectar-feeding birds that rely on fruit and flowers.

860 - The rainforest is a natural pharmacy, with many plants providing chemicals studied for medicines.

861 - Orchid bees pollinate many tropical orchid flowers in rainforests.

862 - The rainforest's diversity allows animals to fill many roles, including predator, prey, pollinator, and decomposer.

863 - Some rainforest animals have specialized diets that link them to certain plant species, shaping the forest's composition.

864 - The rainforest's constant moisture enables rapid decomposition, returning nutrients to the soil quickly.

865 - A storm can topple trees and create new microhabitats by rearranging the forest.

866 - The rainforest's tall trees create mossy trunks and crevices that host many life forms.

867 - Large rainforest mammals, where present, such as gorillas or elephants, help shape the forest by feeding on plants and dispersing seeds.

868 - In many tropical forests, amphibian populations are climate-sensitive and can rapidly decline over small changes.

869 - Some rainforest birds migrate seasonally to track ripening fruit.

870 - The hoatzin is a unique rainforest bird with wing claws in chicks used for climbing.

871 - The harpy eagle nests high in the canopy and can live for decades in its territory.

872 - The jaguar often uses waterways to ambush prey.

873 - The tapir's flexible trunk helps pick leaves and fruit from branches.

874 - The caiman's eyes and nostrils allow it to breathe while mostly submerged.

875 - Some Amazonian fish can breathe air using specialized organs.

876 - The rainforest's many vines, called lianas, act as bridges for animals to move through the canopy.

877 - Fungi in the rainforest help recycle dead leaves and wood into nutrients for plants.

878 - Many rainforest frogs depend on clean, steady rainfall to complete their life cycles.

879 - The forest's microhabitats vary from sunlit gaps to dark shade patches, supporting many species.

880 - The rainforest is a dynamic place where storms create openings that let light reach new plants.

881 - The leaf insect and walking stick mimic the color and shape of leaves to avoid predators.

882 - The electric eel uses specialized organs to generate electricity for hunting and navigation.

883 - The Amazon's pink river dolphins navigate using echolocation to hunt and navigate.

884 - Scent marking by many rainforest mammals helps define territories and communicate with rivals.

885 - Animals such as toucans and monkeys help disperse seeds far from the parent tree.

886 - The rainforest hosts thousands of orchid species, many with unique shapes to lure specialized pollinators.

887 - Epiphytic orchids and bromeliads create pockets of biodiversity high in the trees.

888 - No two square meters of rainforest are exactly the same in species composition, creating endless variety.

889 - Rainforest nights are filled with calls, roars, and clicks that help animals find mates or defend territory.

890 - Some rainforest species rely on mutualism with fungi to digest tough plant material.

891 - Deforestation, habitat fragmentation, and climate change threaten many rainforest species and networks.

892 - The rainforest's layered structure, vibrant colors, and surprising partnerships make it one of Earth's most amazing habitats.

11

DESERT SURVIVORS: LIFE WITHOUT MUCH WATER

Deserts can be scorching hot, bitterly cold at night, and incredibly dry, yet many animals thrive there anyway. In this chapter, you'll discover the clever body tricks and smart behaviors that help desert creatures stay cool, save water, and survive where it almost never rains.

893 - Camels can drink large amounts of water quickly when it is available.

894 - Camel humps store fat, not water, which provides energy and helps generate metabolic water.

895 - Camels regulate body temperature to avoid sweating.

896 - Camels can close nostrils during sandstorms to keep dust out and moisture in.

897 - Camels' feet are broad to walk on soft sand and prevent sinking.

898 - Camels conserve water by sweating slowly and breathing in careful ways.

899 - Camels can survive days without water.

900 - Kangaroo rats have extremely efficient kidneys producing concentrated urine.

901 - Kangaroo rats stay in cool burrows by day and forage at night.

902 - Fennec foxes derive moisture from their prey and often do not need to drink water.

903 - Fennec foxes use their ears for excellent hearing to locate prey underground.

904 - Meerkats live in burrow networks that stay cool in hot weather.

905 - Meerkats take turns on lookout to protect the group while others forage.

906 - Meerkats can go without free water by deriving moisture from prey.

907 - Namib Desert beetles collect water from fog on their backs.

908 - Water droplets condense on Namib beetle's back and flow to its mouth.

909 - Namib beetle shell patterns help droplets form and move toward its mouth.

910 - Saharan silver ants tolerate extreme heat on the surface of the desert.

911 - They have reflective scales that help keep their bodies cool in the sun.

912 - Long legs keep their bodies elevated on hot sand to reduce heat absorption.

913 - The sandfish skink can swim through loose sand to avoid heat.

914 - Its slim body and smooth scales minimize heat absorption when moving underground.

915 - Desert tortoises spend hot days in shade or burrows to stay cool.

916 - Desert tortoises drink water when available and store it in their bodies for later use.

917 - Tortoises have thick skin and shells that minimize water loss.

918 - Gila monsters store fat in their tails and can survive without water for long stretches.

919 - Gila monsters obtain most of their water from prey rather than standing water.

920 - Gila monsters move slowly to conserve water and energy.

921 - Sidewinder rattlesnakes move sideways to minimize contact with hot sand.

922 - Sidewinder rattlesnakes have heat-sensing pits to detect warm prey in the desert.

923 - Desert scorpions glow under ultraviolet light, helping researchers spot them at night.

924 - Addax antelopes have pale coats reflecting sunlight to stay cooler.

925 - Addax can survive on little water by deriving moisture from the plants they eat.

926 - Arabian oryx can tolerate high salt content and filter out salt with specialized glands and kidneys.

927 - Oryx have efficient kidneys to conserve water under extreme heat.

928 - Dromedary camels have broad feet for walking on desert sand.

929 - Bactrian camels have two humps for fat storage and energy reserves.

930 - Camel noses help humidify air and reclaim moisture from exhaled breath.

931 - Sandgrouse birds soak their belly feathers to carry water back to their chicks.

932 - Desert iguanas bask in the sun to warm up and retreat to shade to cool down.

933 - Thorny devil lizard channels dew collected on its skin into its mouth for hydration.

934 - Sand-swimmer lizards move under the sand to stay cool and avoid the heat of the surface.

935 - Mojave rattlesnakes regulate temperature through basking and shade.

936 - Many desert reptiles excrete uric acid to conserve water.

937 - Spadefoot toads burrow and rehydrate after rainfall.

938 - Spadefoot toads breed quickly after rainfall to take advantage of new moisture.

939 - Meerkat burrows help maintain stable humidity and cooler conditions.

940 - Burrows reduce water loss by sheltering from wind and sun.

941 - Some desert animals time activity with dawn and dusk to avoid peak heat for cooler foraging.

942 - Nocturnal activity helps conserve water in many species.

943 - Some animals drink dew collected from the ground at night when humidity condenses.

944 - The desert's extreme dryness makes water a precious resource and shapes many survival strategies.

945 - After rainfall, deserts can bloom and support many life forms, creating short-lived but rich habitats.

946 - The Namib fog is a lifeline for many desert organisms, delivering precious moisture.

947 - Many desert animals rely on microhabitats like rock crevices to escape the hottest sun.

948 - Heat-tolerant animals use behavioral cooling like shade-seeking and burrowing during peak heat.

949 - Light-colored fur or scales reflect sunlight and reduce heat load on the body.

950 - Some desert creatures excrete salts through specialized glands to manage salt balance without wasting water.

951 - Metabolic water produced from fat helps water balance in desert animals by creating additional moisture as fat is metabolized.

952 - Kidneys reabsorb water efficiently to minimize loss and keep hydration steady.

953 - Fat and protein breakdown in the body produces water that helps desert animals stay hydrated.

954 - Rainfall events create temporary water sources that trigger mass movement and breeding in opportunistic species.

955 - Desert species adapt to living with limited fresh water each day and find ways to maximize every drop.

956 - A few species can derive water from humidity in the air during cool nights and foggy mornings.

957 - The presence of water often determines animal distribution in deserts, guiding where different species live and move.

958 - Some desert animals have efficient nasal passages to recover water from inhaled air and cool air before it reaches the lungs.

959 - The ground and rock surfaces can harbor cooler microclimates during heat waves, offering shelter and hydration opportunities.

960 - Many desert animals use scent marking to claim territories without needing to travel far in search of water.

961 - The desert supports high biodiversity in microhabitats, from rocky outcrops to dune crests, each with unique water resources.

962 - Desert landscapes like canyons and dune fields create refuges that help animals balance heat, shelter, and limited water.

963 - Desert animals require only small amounts of water to survive daily, relying on efficiency and metabolism.

964 - Heat stress reduces water needs because respiration rates adapt to minimize moisture loss.

965 - Some species survive drought by reducing body mass and metabolic rate to lower water requirements.

966 - The desert's nighttime air can be cooler and sometimes more humid than the daytime air, aiding respiration for active animals.

967 - The desert's extreme dryness drives many animals to develop heightened senses to locate rare water sources.

968 - Some desert animals rely on nocturnal pollinators that drink dew and extract nectar after sundown.

969 - Desert biodiversity includes many tiny, highly efficient water users that cope with limited hydration.

970 - Desert survival requires careful balance of food, water, and shelter resources to endure long dry spells.

971 - The desert contains landscapes with diverse microclimates, supporting a wide range of life adapted to water scarcity.

972 - Some animals store fat and, through metabolism, generate metabolic water to sustain themselves when water is scarce.

973 - Being able to survive on little water lets many desert animals travel long distances in search of moisture.

974 - The harsh desert environment drives remarkable physiological changes in its inhabitants, enabling survival under extreme thirst.

975 - Desert survival is a testament to how life adapts to extreme conditions.

976 - Estivation helps some desert reptiles and mammals survive the hottest, driest months by slowing metabolism and cutting water loss.

977 - Many desert insects wear a waxy coating that dramatically reduces water evaporation in dry heat.

978 - To stay cool and save water, many desert animals shift their activity to the cooler night hours or to the very first light of dawn.

979 - Some animals drink tiny droplets of dew from plant leaves by licking them or rubbing the nose on surfaces where dew gathers.

980 - After rare desert rains, certain species breed quickly and take advantage of the temporary pools before they disappear.

981 - Fat stores can be metabolized to create metabolic water, helping animals stay hydrated during long dry spells.

982 - Microhabitats such as shaded rock crevices, damp sand pockets, and burrow networks act as reliable water sources and refuge from the heat.

983 - Some desert creatures can tolerate higher salt concentrations in their bodies, thanks to efficient kidneys and specialized salt-excreting glands.

12

FROZEN FRONTIERS: ARCTIC AND ANTARCTIC ANIMALS

Welcome to the Frozen Frontiers, where Arctic and Antarctic animals thrive in a world of ice, wind, and water. This chapter reveals how blubber, fur, and clever tricks help creatures hunt, travel, and raise their young on the coldest seas and shores.

984 - Polar bears rely on sea ice to hunt seals.

985 - Polar bears have black skin under translucent fur.

986 - Polar bear fur is made of clear hairs that reflect light.

987 - Arctic fox fur is thick and fluffy to trap heat.

988 - Arctic hares grow white winter fur to blend with snow.

989 - Musk oxen wear a heavy two-layer coat to stay warm in storms.

990 - Musk oxen form protective circles around calves when predators approach.

991 - Walruses use their long tusks to haul themselves onto ice.

992 - Walrus tusks are used in social displays and for defense.

993 - Ringed seals give birth in snow lairs on the ice.

994 - Ringed seals stay close to the ice edge to breathe and hunt.

995 - Bearded seals have long whiskers that help sense prey in the dark.

996 - Narwhals have a long tusk that usually grows from the left canine.

997 - Narwhal tusks can grow quite long and are often used for social signaling.

998 - Beluga whales are famous for singing and a wide range of sounds.

999 - Belugas use echolocation to locate prey and navigate under ice.

1000 - Arctic wolves hunt in packs to catch larger prey.

1001 - Snowy owls hunt lemmings and other small animals in open Arctic landscapes.

1002 - Snowy owls have feathered legs and big eyes for hunting in the snow.

1003 - Lemmings experience boom-and-bust cycles that influence predators.

1004 - Caribou migrate across the Arctic tundra in large herds.

1005 - The Arctic sea hosts many seals that rest on the ice between dives.

1006 - Emperor penguins breed on Antarctic sea ice during winter.

1007 - Emperor penguin males incubate eggs on their feet with a brood pouch.

1008 - Emperor penguin chicks stay in dense groups to keep warm.

1009 - Adélie penguins nest on rocky shores and use pebbles to build nests.

1010 - Adélie penguins feed on krill and small fish.

1011 - Gentoo penguins are among the fastest penguins in the water.

1012 - Chinstrap penguins have a distinct thin black line under the chin.

1013 - Penguins cannot fly but are excellent divers.

1014 - Crabeater seals filter-feed on krill using specialized teeth.

1015 - Weddell seals can dive deep and stay under the ice for long periods.

1016 - Southern elephant seals are among the largest seals and dive deeply for prey.

1017 - Orcas hunt in pods and are top predators in polar seas.

1018 - Orcas in polar waters use coordinated moves to catch prey.

1019 - Krill swarms form the base of the Antarctic food web.

1020 - Antarctic krill feed on phytoplankton in the water.

1021 - Albatrosses glide long distances over the Southern Ocean.

1022 - Skuas raid penguin colonies for eggs and chicks.

1023 - The Southern Ocean supports a rich and diverse polar ecosystem.

1024 - Antarctic fish survive the cold with antifreeze proteins in their blood.

1025 - Notothenioid fishes have special adaptations for icy seas.

1026 - Penguins maintain waterproof feathers thanks to preen oil from a gland near the tail.

1027 - Penguins molt their old feathers to stay waterproof and warm.

1028 - Blubber stores energy and provides insulation for polar mammals.

1029 - Fat acts as a cushion and helps buoyancy for marine mammals.

1030 - The Arctic has long summer days with continuous daylight.

1031 - The Arctic winter brings long nights that test animal endurance.

1032 - Some polar animals rely on diving skills to access deep food sources.

1033 - Krill are tiny shrimp-like creatures that many larger animals depend on.

1034 - A thick layer of fat helps polar animals stay warm in winter.

1035 - Arctic and Antarctic animals rely on seasonal ice for breeding and feeding.

1036 - Thick fur or feathers plus fat reduces heat loss in extreme cold.

1037 - Penguins can drink seawater and excrete the salt through a special gland.

1038 - The penguin feather system includes water-proofing and insulation via oil glands.

1039 - The Arctic experiences long days in summer and long nights in winter, shaping animal behavior.

1040 - Polarlyce? (placeholder)

1041 - Arctic and Antarctic animals often have seasonal color changes to blend with their icy world.

1042 - Not all polar animals are white explorers; some blend with rocks and water.

1043 - Albatrosses rely on wind currents to travel across oceans for food.

1044 - Krill swarms can be enormous and shift with phytoplankton blooms.

1045 - Krill form a crucial link between microscopic plants and top predators.

1046 - Orcas have sophisticated hunting techniques and social structures.

1047 - Beluga calves learn complex sounds from their mothers.

1048 - Mammals in cold regions often store energy as fat to survive lean times.

1049 - The ice edge is a dynamic frontier where many predators and prey meet.

1050 - Penguin chicks grow waterproof feathers before their first big dives.

1051 - Penguin groups provide warmth through collective behavior during storms.

1052 - Arctic wolves move across vast territories in search of food.

1053 - Bearded seals rest on ice and forage in shallow waters.

1054 - Ringed seals nurse their pups in warm snow lairs on the ice.

1055 - Arctic and Antarctic ecosystems are tightly connected through nutrient cycles and ocean currents.

1056 - Weddell seals are known for creating and maintaining breathing holes in the ice.

1057 - Southern elephant seals perform long deep dives to reach prey in the dark depths.

1058 - Penguin chicks depend on both parents for care during the early weeks after hatching.

1059 - Belugas can change the shape of their melon to communicate underwater.

1060 - Arctic ice loss reduces hunting grounds for many ice-dependent species.

1061 - Antarctic krill swarms can be so dense they appear as brown or pink clouds in the water.

1062 - Penguin colonies are often crowded, with careful social behavior to reduce fights over space and food.

1063 - Sea ice acts as a nursery for young seals and penguins before they venture into open water.

1064 - The life cycles of polar animals are tightly linked to the seasons of ice and water.

1065 - Polynyas are wind-driven or upwelling openings in sea ice that act as tiny underwater oases, concentrating krill, fish, and seals and drawing in predators.

1066 - Antarctic icefish can have transparent blood because they lack hemoglobin, and their bodies compensate with enlarged hearts and gills that extract oxygen efficiently in super-cold water.

1067 - Walruses use their long, sensitive vibrissae to detect movements and vibrations of prey on the seafloor beneath the ice.

1068 - Polar bears can fast for months, surviving on fat reserves while waiting for thick sea ice to return so they can hunt seals again.

1069 - Arctic foxes grow a two-layer coat with a warm underfur and a water-repellent outer layer that helps them stay warm even in brutal winds.

1070 - Orca pods in polar regions often have their own dialects, with calves learning the group's unique calls and hunting tricks from relatives.

1071 - Antarctic krill blooms respond to daylight and phytoplankton productivity, creating shifting feeding carpets that drive the entire polar food web.

13

GRASSLANDS AND SAVANNAS: THE WORLD OF HERDS AND HUNTERS

In the grasslands and savannas, mighty herds roam and agile hunters prowl. These open landscapes host spectacular migrations, clever defenses, and fast, dramatic chases. Read on to discover how animals survive in big, windy spaces.

1072 - Vast grasslands are dominated by grasses with scattered trees that provide shade and shelter.

1073 - Seasonal rains trigger fresh growth, guiding the timing of migrations.

1074 - The Serengeti-Mara ecosystem hosts one of the planet's most famous migrations.

1075 - Wildebeest often move with zebras and gazelles as they follow the rains.

1076 - Zebras' stripes may help break up their silhouette in tall grasses and confuse predators.

1077 - Impalas use agile leaps and fast turns to escape predators.

1078 - Antelope herds rely on many eyes and ears to detect danger early.

1079 - Lions typically hunt in prides that cooperate to capture prey.

1080 - Lions may also scavenge and steal kills from other predators when possible.

1081 - Cheetahs rely on unmatched bursts of speed to outrun most prey.

1082 - Cheetahs prefer open spaces where they can spot grazing animals from far away.

1083 - Leopards hide in trees or tall grass to ambush prey and stay safe from bigger predators.

1084 - Leopards may haul kills into trees to prevent scavengers from stealing them.

1085 - Hyenas hunt in packs and can chase prey for long distances through the grass.

1086 - African wild dogs hunt in coordinated packs with teamwork that improves hunting success.

1087 - Wild dog packs communicate with a mix of vocalizations and body language while chasing.

1088 - Elephants move in matriarchal family groups that remember water sources across years.

1089 - Elephant herds travel to find reliable water and patchy forage during dry seasons.

1090 - Elephants can create paths, wallows, and shade that benefit many other animals.

1091 - Elephants use their trunks to pick up water and food and to smell distant scents.

1092 - Rhinos have thick skin and large horns that help defend against threats.

1093 - Secretary birds hunt snakes by stomping and striking from tall grass.

1094 - Kori bustards are among Africa's largest ground-dwelling birds and move through the grass.

1095 - Ground-nesting birds like bustards use camouflage and quick takeoffs to avoid danger.

1096 - The savanna hosts a huge variety of insects that feed grazing animals, from beetles to grasshoppers.

1097 - Fire plays a natural role in many grassland ecosystems, renewing growth and maintaining balance.

1098 - After fires, new grasses and forbs quickly sprout, feeding grazing animals.

1099 - Prairie dogs of North America live in elaborate underground towns with sentinels.

1100 - Bison herds move seasonally to find fresh grasses and water across the plains.

1101 - Bison can reach surprising speeds when they need to escape danger.

1102 - Bison wallows create muddy spots that animals use to cool down.

1103 - South America's pampas are vast grasslands home to grazing animals like rheas and armadillos.

1104 - Rheas are large flightless birds that run fast to escape predators.

1105 - Armadillos and other small mammals inhabit grasslands and help aerate the soil.

1106 - The Acacia tree common in African savannas provides food, shade, and shelter for many species.

1107 - Acacia trees host mutualistic relationships with ants that defend the tree from herbivores.

1108 - Deep-rooted grasses survive drought by accessing underground moisture.

1109 - Grazing is important for cycling nutrients and keeping plant communities balanced.

1110 - Many grasses recover quickly after being grazed or burned.

1111 - The dry season concentrates herbivores around water sources, changing predator-prey dynamics.

1112 - Predators often target the youngest, smallest, or slowest prey when hunting.

1113 - Herbivores rely on collective vigilance and rapid movement to escape threats.

1114 - Waterholes become bustling meeting places where many species interact.

1115 - The grasslands' winds and dust can help or hinder vision during hunts.

1116 - Early morning light is a prime time for many savanna predators to hunt.

1117 - Vision is crucial for many animals in detecting distant silhouettes in the grass.

1118 - The savanna supports a wide variety of reptiles, including snakes and lizards.

1119 - Birds of prey rise on thermals to scan wide grasslands for movement.

1120 - In some regions, small mammals hide in burrows to escape the heat.

1121 - Prairie dogs are true ecosystem engineers, shaping plant communities with their digging.

1122 - Large herbivores can influence the distribution of grasses by their feeding patterns.

1123 - Grasslands provide important pollinator habitats for bees and butterflies.

1124 - Grassland soils often stay warm at night, helping microbial activity resume quickly.

1125 - Ranchers and wildlife managers work to balance grazing and conservation in grassland areas.

1126 - The appearance of green shoots after rains signals new life across the plains.

1127 - Some grassland birds migrate seasonally between areas with different food resources.

1128 - The speed and agility of grassland predators help keep prey populations in check.

1129 - Antelope species use a combination of speed and herding behavior to survive.

1130 - The Serengeti's seasonal rainfall patterns drive many herd movements.

1131 - The ground in grasslands is sometimes dotted with termite mounds that support wildlife.

1132 - Termites create nutrient-rich soils that benefit grasses and other plants.

1133 - Some grasses have sharp leaves or tough fibers to resist grazing.

1134 - The open landscape helps prey animals detect danger from a distance.

1135 - Predators use wind direction to mask their approach when stalking prey.

1136 - Large migrations create vast dust plumes visible from afar.

1137 - Birds such as vultures and jackals follow predators to feed on leftovers.

1138 - The savanna climate includes distinct wet and dry seasons that shape behavior.

1139 - Camouflage helps many animals remain hidden from predators in tall grass.

1140 - Ground-dwelling birds such as partridges and quails may hide in grasses during the day.

1141 - The life in grasslands depends on the health of the soil and its microbial community.

1142 - Many grazing animals are social and rely on group movements to survive.

1143 - Grasslands across continents show convergent features despite different animals.

1144 - Pronghorns of North America are among the fastest runners on the plains.

1145 - Prairie lizards bask in the sun to warm up before foraging.

1146 - Coyotes and foxes adapt to open spaces with keen senses and versatile diets.

1147 - Jackals are opportunistic scavengers and predators in some savannas.

1148 - A successful chase on the savanna often ends with a quick dash for cover rather than a meal.

1149 - Plants in grasslands rely on broad root systems to recover after grazing.

1150 - The interplay between grazing and fire helps maintain a healthy balance of grasses and forbs.

1151 - Some grassland animals drink from seasonal waterholes that appear after rains.

1152 - Termites and ants help create microhabitats in grasslands and feed soil organisms.

1153 - Some grasses release seeds after fire, helping the plant community regenerate quickly.

1154 - The savanna mosaic supports many species with different lifestyles, from grazers to ambush hunters.

1155 - Animal horns and antlers are used in displays, fights, and defense.

1156 - Many animals have acute hearing and scenting abilities to detect distant danger.

1157 - Antelopes have a variety of horn shapes and sizes that help in defense and competition.

1158 - Predation and prey interactions help keep grassland ecosystems balanced.

1159 - The savanna challenges survival with heat, drought, and predators.

1160 - Grasslands host thousands of species, each adapted to open-space life.

1161 - Migration seasons bring dramatic movements, competition, and opportunities for feeding.

1162 - The open landscape rewards speed, teamwork, and quick thinking when threats appear.

1163 - Some species migrate to escape cold weather at higher latitudes.

1164 - Grasses store carbon in their deep root systems and help renew healthy soils.

1165 - Grasslands and savannas tell stories of speed, teamwork, and survival.

14

RIVERS, LAKES, AND WETLANDS: FRESHWATER SUPERSTARS

Rivers, lakes, and wetlands are full of freshwater superstars—the animals that make these waters pulse with life. From beavers building busy dams to otters skimming through ripples, these habitats host extraordinary tricks for feeding, staying safe, and surviving.

1166 - Freshwater amphibians begin life in water as eggs without shells, then hatch into gilled larvae before metamorphosing.

1167 - The axolotl stays aquatic and keeps gills for life, even as an adult.

1168 - Wood frogs can freeze solid in winter and then thaw and move again in spring.

1169 - Salamanders can breathe through their skin and lungs, depending on the species.

1170 - Newts can release potent toxins from skin glands when threatened.

1171 - Some frogs guard eggs or carry tadpoles on their bodies.

1172 - Tadpoles can be herbivores, omnivores, or carnivores depending on species.

1173 - Spotted salamanders spend their lives in wetlands and breed in ponds.

1174 - The common toad often dwells in damp soils and emerges mainly after rains.

1175 - Amphibians are excellent indicators of wetland health because their skin absorbs chemicals.

1176 - Frogs use vocal sacs to amplify their mating calls.

1177 - The goliath frog can reach large sizes in African streams.

1178 - Amphibians are among the earliest land animals and rely on fresh water for reproduction.

1179 - The hellbender is a huge salamander living in fast North American streams.

1180 - The rough-skinned newt is one of the most toxic amphibians in North America.

1181 - Freshwater amphibians face threats from pollution, habitat loss, and climate change.

1182 - Crocodilians include alligators, crocodiles, caimans, and gharials.

1183 - They live mainly in freshwater habitats such as rivers, lakes, and swamps, with some tolerating brackish water.

1184 - They can hold their breath for up to two hours by slowing their heart rate.

1185 - They have a nictitating membrane, a transparent third eyelid, to protect their eyes underwater.

1186 - They have powerful jaws that can grab prey and deliver a strong bite.

1187 - They lay eggs in nests and temperature determines the sex of hatchlings.

1188 - Mothers often guard nests and young after hatching.

1189 - They are apex predators that shape the communities around wetlands.

1190 - Their bodies are covered with scaly armor called scutes.

1191 - They swim with strong, muscular tails that propel them through water.

1192 - They have sensitive pressure receptors on their jaws to detect movement in water.

1193 - The gharial has a very long, slender snout specialized for catching fish.

1194 - Caimans vary in size and habitat from marshy swamps to slow rivers.

1195 - Alligators are mostly freshwater, while some crocodile species venture into saltwater.

1196 - They communicate with bellows, hisses, and social sounds to attract mates and defend territory.

1197 - They lay eggs in nests of vegetation and soil and guard them.

1198 - They protect and guide their young for a time after hatching.

1199 - They can perform a dramatic death roll to subdue large prey.

1200 - Beavers are among the largest rodents in North America and Europe.

1201 - They build dams to slow water flow and create ponds.

1202 - They build lodges with underwater entrances for safety.

1203 - Their incisor teeth grow continuously and have orange color due to iron.

1204 - They have webbed hind feet for swimming.

1205 - They use their tails as a paddle and signaling device.

1206 - They live in family groups led by a mating pair.

1207 - They store food underwater near their dam for winter.

1208 - They scent-mark their territory with castoreum.

1209 - Their damming reshapes rivers and creates habitats for many species.

1210 - They rely on wetlands for food and shelter.

1211 - They can hold breath for up to 15 minutes underwater.

1212 - They are herbivores feeding on bark, twigs, and aquatic plants.

1213 - They chew through trees using their strong jaws.

1214 - They use clever engineering to manage water flow.

1215 - River otters have extremely dense fur that keeps them warm in cold water.

1216 - They have long whiskers that help detect vibrations in water.

1217 - They can close their nostrils and ears when diving.

1218 - They are playful and often hold hands to form rafts when they sleep.

1219 - They use their sharp teeth to catch fish and other prey.

1220 - They are excellent swimmers with webbed feet.

1221 - They groom their fur to maintain its water-proofing.

1222 - They have a varied diet including fish, crayfish, frogs, and small invertebrates.

1223 - They mark territory with scent.

1224 - They live in dens along riverbanks or along coastal edges.

1225 - They can be social and travel in family groups.

1226 - They communicate with a range of vocalizations.

1227 - They help keep wetlands healthy by controlling prey populations.

1228 - They are found in North America, Europe, and parts of Asia depending on species.

1229 - They are champions of clean, quick river flows.

1230 - They are a favorite of kids for their playful antics.

1231 - They can deliver powerful shocks to stun prey and deter predators.

1232 - Electric eels live in murky, oxygen-poor waters of the Amazon and Orinoco basins.

1233 - They can grow up to about 2.5 meters long.

1234 - They use electricity to navigate and communicate in the dark water.

1235 - Lungfish have both gills and lungs for breathing.

1236 - They can survive droughts by burrowing in mud and entering estivation.

1237 - Some lungfish can live for years out of water during dry seasons.

1238 - Mudskippers spend most of their time on land in mangrove swamps.

1239 - They can breathe through their skin and mouth membranes when out of water.

1240 - Mudskippers can climb mangrove roots and move across the mud.

1241 - They use their fins to walk on land.

1242 - They have eyes that can move independently for better land and water vision.

1243 - Walking catfish can walk across damp ground by wriggling and using their fins.

1244 - They can survive on land for hours to move between water bodies.

1245 - Paddlefish have rostrums filled with sensory organs to detect plankton and electric fields.

1246 - They are filter feeders that strain tiny organisms from the water.

1247 - Sturgeon are ancient fish with scutes along their bodies.

1248 - They migrate long distances to spawn in large rivers.

1249 - Some sturgeon species can live a very long time, sometimes decades.

1250 - Arapaima gigas is one of the largest freshwater fishes and can reach large sizes.

1251 - It can breathe air through a specialized labyrinth organ.

1252 - It surfaces to gulp air to oxygenate its blood.

1253 - Arapaima are known as pirarucu in Brazil.

1254 - Arapaima have tough, overlapping scales for protection.

1255 - Gar are ancient, elongated predatory fish with sharp teeth.

1256 - They can gulp air to breathe when water is low in oxygen.

1257 - Bowfin have a primitive lung-like structure that helps them breathe air.

1258 - Bowfin can live in slow-moving, stagnant waters where other fish struggle.

1259 - Freshwater fish show a wide variety of colors and mating displays to attract partners.

15

CREEPY-COOL INSECTS
AND OTHER ARTHROPODS

Step into the creepy-cool world of insects and other arthropods, where tiny bodies hide extraordinary powers. This chapter dives into metamorphosis, silk, venom, teamwork, and the idea that tiny creatures run much of the planet.

1260 - Insects often transform radically through metamorphosis, turning small larvae into very different-looking adults.

1261 - Some insects, like butterflies, beetles, and bees, experience complete metamorphosis with four life stages: egg, larva, pupa, and adult.

1262 - Other insects such as crickets and true bugs undergo incomplete metamorphosis, where nymphs look like smaller versions of adults.

1263 - Spiders are arachnids, not insects, and many spin silk to build webs, sacs, and safety lines.

1264 - Spider silk is remarkably strong and can hold a heavy prey without breaking.

1265 - Some spiders use silk to balloon, letting young spiderlings drift on air currents to new homes.

1266 - Silk is also used to wrap eggs, construct protective nurseries, and line nests.

1267 - Venom is a chemical weapon used by many arthropods to immobilize prey or deter threats.

1268 - Venom varies widely—some targets the nervous system, others disrupt muscles, and some only mildly affect humans.

1269 - Scorpions glow under ultraviolet light due to special chemicals in their bodies.

1270 - Scorpions can slow their metabolism to survive long stretches without food.

1271 - Some scorpions are among the most venomous land predators, though bites to humans are rare.

1272 - Ants, termites, bees, and wasps are social insects with complex colonies and specialized castes.

1273 - Termites build enormous mounds that regulate temperature and humidity for colony life.

1274 - Ants use pheromones, touch, and sound to communicate and stay organized.

1275 - Some termite species form very large colonies that span large areas.

1276 - The mantis shrimp delivers one of the fastest and most powerful strikes in the ocean.

1277 - Its club-like appendage can slam shut with enough force to crack hard shells.

1278 - Mantis shrimp eyes are incredibly complex, giving rich color vision and depth perception.

1279 - Beetles often have impressive horns, plates, or luminous patterns used in battles and courtship.

1280 - Beetles are among the most diverse groups on Earth, living in nearly every habitat.

1281 - Jumping spiders see with eight eyes and can leap long distances to surprise prey.

1282 - Jumping spiders often perform elaborate courtship displays to avoid mistaken predation.

1283 - Silk has inspired human innovations in fabrics, medical sutures, and biodegradable materials.

1284 - Orb-weaver spiders weave circular webs with strategic spacing to catch flying insects.

1285 - Not all spiders spin webs; some hunt prey directly, like wolf spiders and fishing spiders.

1286 - Tarantulas carry their young on their backs for protection after hatching.

1287 - Scorpions give birth to live young that ride on the mother's back before their first molt.

1288 - Some spiders can survive long droughts by reducing activity and conserving water.

1289 - Crabs and lobsters belong to the crustaceans, a major group of arthropods with hard exoskeletons.

1290 - Crustaceans include crabs, lobsters, shrimp, and barnacles, living in oceans and freshwater.

1291 - The mantis shrimp's punch is so fast it creates a shockwave that can stun prey even in cloudy water.

1292 - Many crustaceans carry eggs attached to the abdomen until they hatch, giving young a ride to safety.

1293 - Millipedes coil into balls or spirals when threatened and release defensive substances.

1294 - Centipedes are fast, venomous predators with one pair of legs per body segment.

1295 - Millipedes typically have two pairs of legs per body segment and feed on decaying plant matter.

1296 - Arthropods rely on molts to grow, shedding their outer skin as they increase in size.

1297 - Molting leaves arthropods temporarily vulnerable while new skins harden.

1298 - Some spiders can regenerate lost legs during subsequent molts.

1299 - Termites and ants recycle wood and plant fiber, releasing nutrients back into ecosystems.

1300 - Bees pollinate many flowering plants, enabling fruit and seed production.

1301 - Bumblebees can fly at cooler temperatures thanks to their large wings and fuzzy bodies.

1302 - Mosquitoes use a long proboscis to pierce skin and drink blood, often without being felt immediately.

1303 - The long antennae of many insects help detect chemical cues, air movements, and touch.

1304 - Some insects mimic dangerous species to avoid predators through deceptive appearance.

1305 - The peppered moth is a famous example of natural selection in action, changing color with pollution levels in habitats.

1306 - Spiders sense vibrations through their webs, allowing rapid responses to prey or danger.

1307 - Insects communicate with pheromones that guide others to food or warn of danger.

1308 - Ants often farm aphids or scale insects, protecting them in exchange for honeydew or other secretions.

1309 - Some ants form living bridges with their bodies to help others cross gaps during foraging.

1310 - Termites rely on gut microbes to digest wood, extracting nutrients that are otherwise hard to access.

1311 - Social insect colonies often include workers, soldiers, and queens to maintain survival and reproduction.

1312 - Fireflies produce light through bioluminescence to attract mates and signal territory.

1313 - Firefly light is produced by a chemical reaction inside specialized light-emitting organs, not by heat.

1314 - Spiders can adjust the thickness and type of silk they spin for different purposes.

1315 - Some spiders wrap prey in silk before delivering a venom bite, helping immobilization.

1316 - Insects such as lacewings and ladybugs help control pests by eating aphids and mites.

1317 - The bee's stinger is a modified ovipositor in female bees, used for defense or predation.

1318 - Some wasps are solitary and hunt caterpillars to provision their nests for their offspring.

1319 - The world's smallest arthropods can occupy tiny microhabitats, from leaf litter to tree bark crevices.

1320 - The eyes of many arachnids are simple and specialized for detecting movement in dim light.

1321 - The pincer-like mouthparts of some arachnids help them grasp prey or defend themselves.

1322 - The jaw-like mandibles of many insects allow slicing, chewing, and carrying prey.

1323 - Bioluminescence in some crustaceans and other marine arthropods helps attract mates and confuse predators in the deep sea.

1324 - Some spiders lay eggs in silk sacs that they guard until the young hatch.

1325 - Termite mounds can create microhabitats that support a surprising diversity of life.

1326 - Ant colonies can store huge amounts of food in their nests through careful organization and foraging.

1327 - Bees use a combination of dancing, scent, and vibration to communicate complex foraging information.

1328 - The old Earth was home to giant arthropods in prehistoric times, showing how diverse this group can be.

1329 - Parasitic wasps lay eggs inside other insects, and the larvae feed on the host as they develop.

1330 - Some spiders use ultraviolet markings on webs to attract prey or attract mates.

1331 - Silk threads from spiders can be stronger than steel when comparing equal weight.

1332 - Crabs and lobsters molt to grow, leaving them temporarily vulnerable to predators.

1333 - Mites are tiny arachnids that live everywhere, from skin and hair to soil and plants.

1334 - Some mites feed on plants as pest insects, while others feed on animal hosts.

1335 - The diversity of arthropods is enormous, making up the majority of species on Earth.

1336 - Some arthropods perform remarkable feats of endurance, traveling long distances to find food or mates.

1337 - The gut microbiome of many arthropods helps them digest tough plant material or wood.

1338 - The chemical defenses of some beetles help deter predators with smells or tastes that are unpleasant.

1339 - Ants' underground tunnels can influence soil structure and aeration, benefiting plant roots.

1340 - Some arthropods practice cooperative breeding, with non-breeding individuals helping raise young.

1341 - The wings of many insects are lightweight but provide significant lift for flight.

1342 - The eyes of many crustaceans can detect polarized light, helping navigation and hunting in water.

1343 - The life cycles of aquatic arthropods show incredible adaptation to both water and land environments.

1344 - Some spiders are social and live in colonies with shared nests and resources.

1345 - Certain crustaceans, like shrimp, can live in both freshwater and marine environments.

1346 - Arthropod diversity includes many parasitic species that infect other animals or plants.

1347 - Honeybees have a highly organized social life with division of labor and precise timing in tasks.

1348 - Some insects have specialized mouthparts suited to feeding on nectar, plants, or blood.

1349 - Many arthropods use camouflage to blend into leaves, bark, or soil to avoid predators.

1350 - The ecological impact of arthropods is immense, driving pollination, decomposition, and nutrient cycling.

1351 - The ability to regrow lost limbs in some arthropods demonstrates remarkable regeneration.

1352 - Tiny arthropods form the backbone of many food webs, supporting larger animals from birds to whales.

16

REPTILES AND AMPHIBIANS: SCALES, SLIME, AND SURVIVAL

Cold-blooded creatures like frogs, lizards, turtles, and snakes live surprising lives in many habitats. In this chapter you'll learn about shedding skin, growing back body parts, slime and poison defenses, and the incredible tongue tricks that help frogs and reptiles survive.

1353 - Reptiles and amphibians are ectothermic, meaning their body temperature depends on the environment.

1354 - Most reptiles shed their skin as they grow, often in one piece or in patches.

1355 - Some snakes shed their skin in a single, continuous sheet.

1356 - Many lizards can drop their tails to escape predators, a defense called caudal autotomy.

1357 - After tail loss, many lizards can regrow a new tail, though the replacement may look different.

1358 - Salamanders and newts are famous for regenerating lost limbs, tails, and sometimes other body parts.

1359 - Amphibian skin is moist and permeable, aiding in respiration and moisture exchange.

1360 - Not all frogs are poisonous, and many have harmless skin.

1361 - Venomous snakes inject venom through fangs to immobilize prey.

1362 - Snake venom varies by species; some are mainly neurotoxic, others are mainly hemotoxic.

1363 - Pit vipers have heat-sensing pits that help them detect warm-blooded prey in the dark.

1364 - Frogs begin life as eggs laid in water, which hatch into tadpoles.

1365 - Tadpoles develop legs and lose their tails as they metamorphose into adult frogs.

1366 - Some salamanders begin life in water and mature into terrestrial adults.

1367 - Some frogs undergo direct metamorphosis, hatching as small frogs rather than tadpoles.

1368 - Chameleons catch prey with long, sticky tongues that shoot out rapidly.

1369 - The chameleon tongue can extend beyond its body length to reach distant prey.

1370 - Frogs use their tongues to capture prey with speed and precision.

1371 - Some lizards have strong tails used for defense, balance, and communication.

1372 - Reptiles use color and pattern for camouflage to blend into their environments.

1373 - Turtles have shells made of bone and keratin that protect their bodies.

1374 - Turtle shells help regulate body temperature in many species.

1375 - Some turtles can pull their heads and limbs completely inside their shells for protection.

1376 - Sea turtles migrate long distances across oceans to reach feeding grounds and nesting sites.

1377 - Many turtles can stay underwater for extended periods by slowing their metabolism and holding their breath.

1378 - Amphibians have moist skin that remains permeable, making them sensitive to dryness and pollutants.

1379 - Amphibian eggs generally lack shells and must remain moist to survive.

1380 - Frogs and toads often use vocal calls to attract mates and defend territories.

1381 - Some toads secrete toxins from glands behind their eyes to deter predators.

1382 - Parotoid glands on toads release toxins when threatened.

1383 - Some lizards are venomous, such as Gila monsters and beaded lizards.

1384 - Axolotls are a type of salamander that retains juvenile features into adulthood, a phenomenon called neoteny.

1385 - Tuataras are ancient reptiles from New Zealand, often called living fossils.

1386 - Tuataras have a third eye on top of their heads that helps sense light and regulate daily activity.

1387 - Crocodilians have a four-chambered heart, which supports efficient separation of oxygenated and deoxygenated blood.

1388 - Basilisk lizards can run on the surface of water for short distances thanks to their speed and special foot fringe.

1389 - Flying dragons (Draco lizards) glide between trees using wing-like flaps of skin.

1390 - Salamanders are often active at night and hunt small invertebrates by scent and vibration.

1391 - Frogs can jump remarkable distances with powerful hind legs.

1392 - Snakes move by contracting their muscles and sliding their scales against the ground in a wriggling motion.

1393 - Some snakes can climb trees using their scales, muscles, and body flexibility.

1394 - Reptiles shed old skin and often replace teeth throughout life.

1395 - Amphibians can absorb water through their skin, aiding hydration in damp environments.

1396 - Some frogs lay eggs in clusters that float on the water surface.

1397 - Some frogs lay eggs in strings that dangle in the water.

1398 - Some frogs guard their eggs and tadpoles, while others leave them to develop on their own.

1399 - Many lizards lay eggs, while others give birth to live young.

1400 - Some lizards have colorful dewlaps or crests used in courtship displays.

1401 - Some frogs can survive freezing temperatures by producing antifreeze-like substances in their bodies.

1402 - Salamanders have smooth, moist skin that can secret protective mucus.

1403 - Some frogs breathe through their skin in addition to their lungs.

1404 - Some reptiles have venomous bites that can cause pain, swelling, or more serious effects.

1405 - Some lizards reproduce by parthenogenesis, producing offspring without fertilization in certain species.

1406 - The eyes of many reptiles are protected by a clear, movable structure called a brille, so they don't have move-able eyelids.

1407 - Some frogs can change color to blend into different backgrounds for camouflage or social signaling.

1408 - Some snakes have patterned skins that help them blend into leaves, sand, or water.

1409 - Some frogs breed in temporary pools with rapid metamorphosis to escape drying habitats.

1410 - Some salamanders are fully aquatic throughout life.

1411 - Some frogs have specialized calls to attract mates at specific times or in crowded choruses.

1412 - The teeth of many reptiles grow continuously and are replaced when worn.

1413 - Some snakes use constriction to subdue prey without venom.

1414 - The venom of Gila monsters and beaded lizards is delivered through their fangs.

1415 - Some lizards use scent-marking to defend territory or attract mates.

1416 - The structure of reptile lungs supports efficient breathing during activity and rest.

1417 - Some amphibians secrete toxins through their skin as a chemical defense.

1418 - Some frogs are small but produce loud, far-carrying calls.

1419 - The tail of many lizards can act as a decoy during predator attacks.

1420 - Some salamanders have gills during larval stages, which they lose as they metamorphose.

1421 - Some reptiles enter brumation or hibernation during cold months.

1422 - Amphibians occur in a wide range of habitats, provided there is moisture or water.

1423 - Some reptiles survive in deserts by conserving water and seeking shade and shelter.

1424 - Some amphibians have glands that produce toxins or repellent substances.

1425 - Some lizards shed their tails in pieces during autotomy.

1426 - Frogs have eyes that provide broad field of view for spotting prey and predators.

1427 - Snakes have a forked tongue that helps them sense the environment.

1428 - Some lizards can rapidly change color in response to temperature, mood, or environment.

1429 - Amphibians rely on both aquatic and terrestrial habitats across their life cycles.

1430 - The color of reptile skin can shift with temperature changes due to pigment cell activity.

1431 - Crocodilians are known for powerful bites and efficient predation.

1432 - Many reptiles are adept climbers, using claws and flexible bodies.

1433 - Some amphibians breed in response to rainfall and seasonal changes.

1434 - The metamorphosis of frogs can occur rapidly or gradually depending on species and conditions.

1435 - Many reptiles lay eggs in protected nests, burying them in sand or leaf litter.

1436 - Some amphibians use pheromones and chemical cues to attract mates.

1437 - Some reptiles shed their skin to remove parasites and keep their bodies clean.

1438 - The survival of reptiles and amphibians depends on clean water and abundant prey.

1439 - Reptiles and amphibians rely on camouflage, toxins, venom, metamorphosis, and regeneration to survive in their habitats.

17

BIRD BRILLIANCE: FEATHERS, FLIGHT, AND FANCY FEATS

From hummingbirds that hover to penguins that glide underwater, birds show dazzling feats of flight, navigation, nest-building, mimicry, and display. In this chapter, gentle giants and tiny acrobats alike reveal how feathers power movement, maps guide journeys, nests shelter families, and songs and dances woo mates.

1440 - A bird's wing functions as an airfoil, producing lift as air moves over its curved upper surface.

1441 - The alula is a small feather on the wing that helps control airflow and prevent stalls.

1442 - Birds have hollow bones that keep the skeleton light for flight.

1443 - Many birds switch between flapping and gliding to save energy during long flights.

1444 - Birds breathe with lungs and additional air sacs that improve oxygen delivery during rapid wingbeats.

1445 - The tail acts as a rudder, helping steer and balance in flight.

1446 - In level flight, some birds can reach speeds around 60 miles per hour depending on size and species.

1447 - Hummingbirds can hover precisely by adjusting wing beat and body tilt.

1448 - Some birds can hover for extended periods by balancing lift and body position.

1449 - Birds use wind and air currents to travel longer distances with less effort.

1450 - The shape of a bird's wings affects speed, turning ability, and energy use.

1451 - Some birds sleep with half their brain at a time to stay alert for danger.

1452 - Albatrosses can cover thousands of miles across oceans during migration.

1453 - Pigeons display impressive navigation skills, using multiple cues to find their way home.

1454 - Many birds use the sun as a compass and adjust their flight direction as the sun moves.

1455 - At night, many birds navigate by the patterns of stars in the sky.

1456 - Birds rely on landmarks such as coastlines and rivers to guide their routes.

1457 - Some birds can drink seawater thanks to salt glands that remove excess salt.

1458 - Feathers are made of keratin and provide insulation, waterproofing, and lift.

1459 - Parrots are famous for mimicking human speech and other sounds.

1460 - Mockingbirds and thrashers mimic a wide range of bird calls.

1461 - The lyrebird is a master mimic, able to copy many everyday sounds.

1462 - Bird songs are learned and refined as birds grow, often with help from older birds.

1463 - Courtship displays often include songs, dances, and dramatic aerial maneuvers.

1464 - Some birds perform dazzling aerial displays during courtship seasons.

1465 - The male bowerbird builds a special display area and decorates it with bright objects.

1466 - Birds nest in a variety of places, from tree hollows to bushes to cliff ledges.

1467 - Weavers create intricate hanging nests woven from grasses and fibers.

1468 - Rockhopper penguins use stones to create a nest bowl for eggs.

1469 - Some birds camouflage their nests with leaves and lichens to blend into the surroundings.

1470 - Cuckoos lay their eggs in other birds' nests, relying on foster parents to raise the chicks.

1471 - Some seabirds decorate their nests with lichens to blend into their surroundings.

1472 - Frigatebirds can stay aloft for days by riding rising air currents.

1473 - Swifts spend most of their lives on the wing, only landing to breed.

1474 - Kingfishers dive from perches to catch fish with their sharp beaks.

1475 - Owls have forward-facing eyes that give them excellent depth perception at night.

1476 - Birds communicate using a rich mix of songs, calls, and physical displays.

1477 - The colors of a bird's plumage can indicate age, sex, and health to rivals and mates.

1478 - Some birds display crest feathers and other ornaments to attract mates.

1479 - The beak shape and size often reflect a bird's primary diet and feeding behavior.

1480 - Feathers provide waterproofing through secret oils and a tight feather structure.

1481 - The arrangement of feathers reduces drag and helps produce lift.

1482 - Birds molt and replace feathers regularly to keep flight abilities sharp.

1483 - Some birds have excellent color vision that helps them pick ripe fruits and navigate.

1484 - Birds use scent in some cases to locate nesting sites or food sources.

1485 - The nocturnal owl uses silent flight and keen senses to hunt in darkness.

1486 - The osprey's talons are specialized for gripping slippery fish during a spectacular hunting dive.

1487 - Some birds build nests on cliffs to stay safe from ground predators.

1488 - The beak of a toucan can help with a unique feeding method in the rainforest canopy.

1489 - The kiwi is a flightless yet flightless bird with a long beak and strong senses of smell.

1490 - The cassowary is a large flightless bird that lives in rainforest habitats and uses powerful legs for defense.

1491 - Some birds migrate along coastlines to take advantage of consistent wind and food resources.

1492 - Birds in arid regions may travel long distances in search of water and food.

1493 - Barn swallows catch insects on the wing in rapid, acrobatic flights.

1494 - Eagles and hawks rely on extremely keen vision to spot prey from high above.

1495 - The red-tailed hawk uses a high-pitched cry to communicate across territories.

1496 - Some birds produce nonvocal sounds by beating their wings or tapping with their beaks for communication.

1497 - Birds can have complex social structures that help in defense and foraging.

1498 - Some parrots can learn to imitate environmental noises like alarms or camera shutters.

1499 - Birds can coordinate in groups for foraging or defense.

1500 - The shape and color of feathers can help conceal a bird when it sits still.

1501 - Some birds nest in colonies to increase protection against predators.

1502 - The Arctic tern migrates incredibly far, traveling from Arctic breeding grounds to Antarctic winters.

1503 - The common swift is renowned for staying aloft for long stretches during its life.

1504 - The horned puffin's colorful beak stands out during breeding season.

1505 - Birds use diverse techniques to keep eggs and chicks warm and safe.

1506 - Young birds learn essential flight skills from parents through guided practice.

1507 - Some birds display in elaborate dances and postures to signal readiness to mate.

1508 - The duck's webbed feet are adapted for efficient swimming and steering in water.

1509 - Some birds use specialized calls to signal dangers at different distances or elevations.

1510 - Birds often form feeding flocks that help locate food across large areas.

1511 - The sun and wind influence the timing and path of many migratory journeys.

1512 - The night sky provides important cues for navigation for many nocturnal species.

1513 - Some birds use magnetic cues to orient themselves across long journeys.

1514 - The long migrations of some species require stopping at feeding grounds to refuel.

1515 - Birds can adapt their nesting strategies to changing climates and habitats.

1516 - Different species communicate with distinct songs to identify themselves to others of the same species.

1517 - Birds' nests can vary in complexity from simple scrapes to intricate woven structures.

1518 - Young birds depend on parental feeding for a period after hatching.

1519 - The blue-footed booby uses distinctive courtship displays to attract mates.

1520 - Birds' plumage may reflect UV patterns visible to other birds but not to humans.

1521 - The albatross's long journey is supported by oceanic winds and efficient respiration during flight.

1522 - Bird flight is a product of anatomy, energy management, and environmental adaptation.

1523 - Their songs and calls can convey information about territory, distress, or mating readiness.

1524 - Feather color and pattern help birds adapt to their habitats via camouflage or signaling.

1525 - Different birds have evolved a wide range of nest-building techniques to suit their environments.

1526 - Bird flight, navigation, nesting, mimicry, and display collectively illustrate the ingenuity of avian life.

18

MAMMAL MARVELS: FUR, POUCHES, AND INCREDIBLE PARENTING

From fluffy fur to living baby pockets, mammals show amazing ways to care for their young. In this chapter, we explore fur, pouches, and parenting tricks that help mammal babies grow up safe, strong, and curious.

1527 - Fur provides insulation by trapping a layer of air next to the skin.

1528 - Some mammals grow thicker winter coats and shed them as seasons change.

1529 - Fur color and patterns help animals blend in or signal others during mating.

1530 - Whiskers are highly sensitive hairs that help animals sense touch and navigate in the dark.

1531 - Hair grows from follicles all over the body and can renew throughout life.

1532 - Platypus and echidna are the only living mammals that lay eggs.

1533 - Monotremes lack nipples; milk oozes onto the skin for babies to lick.

1534 - Platypus mothers incubate eggs in a burrow for about 10 days before they hatch.

1535 - Echidnas lay eggs that hatch into small, helpless puggles which then nurse.

1536 - Marsupials give birth to very underdeveloped young that finish growing in a pouch.

1537 - The pouch is a warm, protective space where a joey attaches to a teat and nurses.

1538 - Embryonic diapause lets some marsupials pause the development of a new embryo until the current joey is ready to leave the pouch.

1539 - A newborn joey is tiny, often the size of a jelly bean, and must crawl into the pouch on its own.

1540 - Pouches differ in size, shape, and opening across different marsupial species.

1541 - Kangaroos can nurse a joey in the pouch while another embryo waits to develop.

1542 - Koalas carry their young in the pouch for several months as they feed on leaves.

1543 - Wombats have backward-opening pouches to protect the young while digging.

1544 - Some marsupials can nurse more than one joey at a time.

1545 - The early life of most marsupials is spent in the safety of the pouch.

1546 - Mammary glands produce milk that nourishes offspring after birth.

1547 - Milk composition differs between species to meet the growth rates and dietary needs of the young.

1548 - Elephant calves nurse for many years and stay close to their mothers as they learn social behavior.

1549 - Dolphins nurse their calves while the mothers swim and teach them to communicate and swim.

1550 - Sea otter mothers cradle pups on their chests and groom them to keep them warm and clean.

1551 - Beavers raise kits with help from both parents as they build lodges and protect their families.

1552 - Bears care for cubs for many months and teach them to forage and survive in the wild.

1553 - Wolves train pups in hunting, howling, and social rules within the pack.

1554 - Lions teach cubs to stalk and cooperate within the pride.

1555 - Tigers practice hunting with their cubs in safe, concealed areas before bringing them prey.

1556 - Seals nurture pups after birth and guide them to learn swimming.

1557 - Sea lions teach pups to swim and to navigate waves and beaches.

1558 - Whales nurse calves at sea and help them learn songs and social behavior.

1559 - Cheetahs balance nursing with hunting, teaching cubs to be fast and cautious.

1560 - Rabbits nurse newborn kits in sheltered nests and protect them from predators.

1561 - Porcupines protect their young with spines and careful parental care.

1562 - Ferret kits rely on their mother's milk and learn to explore their environment.

1563 - Primates often invest many years in a single offspring, providing food, protection, and social learning.

1564 - Alloparenting is common in many species, with siblings and relatives helping care for the young.

1565 - A mother's milk provides antibodies and nutrients to protect babies from disease.

1566 - Milk contains fats, proteins, and sugars tailored to support growth at different ages.

1567 - The growth of mammal young depends heavily on the energy available from milk.

1568 - Lactation is one of the most energy-intensive activities a mammal performs.

1569 - Some mothers adjust the timing of births based on food availability and season.

1570 - In many species, mothers emit protective calls or alarms to shield their young.

1571 - Dense fur helps aquatic mammals stay warm when they swim.

1572 - The density and structure of fur influence thermoregulation and buoyancy.

1573 - Scent marking and cues help mothers and young locate each other in dense habitats.

1574 - Fathers and other relatives often help with childcare in many mammal societies.

1575 - The timing of birth and lactation often aligns with periods of abundant food.

1576 - Babies learn basic behaviors by observing and imitating their mothers.

1577 - The mother's milk timing and feeding schedule shape early growth patterns.

1578 - Many mammal mothers form strong, lasting bonds with their offspring.

1579 - Some young mammals are born blind and deaf and rely on touch to find their mother.

1580 - Some pouches are lined with specialized tissue to keep the young warm and snug.

1581 - The pace of early development is guided by how quickly milk and nutrition allow growth.

1582 - The lactation period is a major factor in a species' life history and reproduction.

1583 - In some species, fathers guard dens and protect nursing young from predators.

1584 - The fur of arctic animals often includes hollow hairs that trap air for insulation.

1585 - Fur density can adapt in response to climate and altitude over generations.

1586 - Fur can be shed seasonally to adjust insulation as environments change.

1587 - Adult mammals communicate with young through a range of calls, grooming, and touch.

1588 - Calves, cubs, and pups learn to respond to danger cues from their mothers and elders.

1589 - The composition of milk can shift as the young grows.

1590 - The pace of development varies by species, from quick to years-long growth.

1591 - Social structures shape how care for the young is shared within families.

1592 - A well-fed mother can regulate body temperature more easily, aiding lactation.

1593 - Pups often begin exploring with the mother's guidance while still nursing.

1594 - Kangaroo mothers can stay mobile while their joeys nurse in the pouch.

1595 - Koalas spend much of their lives perched high in trees, keeping their young close.

1596 - Many newborn mammals are born undeveloped and rely on sensory cues to locate the mother's nipple.

1597 - Babies use smell to locate their mother's nipple and latch on.

1598 - Milk production is controlled by hormones and adapts to the needs of the growing young.

1599 - In some species, lactation continues even after the young begin eating solids.

1600 - Maternal stress can affect the growth and survival of young.

1601 - Family bonds and social learning often shape future behavior and success.

1602 - Play and social interaction help young mammals learn important survival skills.

1603 - Some mammal species practice cooperative breeding, where non-parent animals help raise young.

1604 - Fur serves multiple roles: insulation, camouflage, signaling, and even waterproofing.

1605 - The microstructure of fur influences warmth and wetness, not just length.

1606 - Variety in dentition relates to diet and lactation stage.

1607 - Lactation is highly energy-demanding, so mothers need nutrient-rich food during nursing.

1608 - Some marine mammals feed their young at the surface, while others dive to forage with calves.

1609 - Certain milks are especially rich to support rapid brain development in pups.

1610 - Offspring survival often hinges on protection from predators and harsh weather.

1611 - The maternal bond with young helps regulate stress and supports development.

1612 - In some species, fathers stay with young for extended periods to help raise them.

1613 - A stable family unit improves reproductive success and thriving.

1614 - Fur, pouches, and parenting strategies illustrate how mammals adapt to diverse habitats.

1615 - Differences in fur and caregiving show the vast diversity of mammals.

1616 - The mother–offspring bond influences social status.

1617 - Playful exploration helps young mammals develop motor and social skills.

1618 - Many species have juvenile-dependent periods that last years.

1619 - Long lactation helps offspring cope with scarce resources.

1620 - Marsupial pouches demonstrate a unique reproductive strategy that adapts to the environment.

1621 - The combination of fur, pouches, and devoted
caregiving defines a core aspect of mammal life.

PREDATORS AND PREY: HUNTING, ESCAPING, AND STAYING ALIVE

Predators chase their prey with skill and speed, while prey develop clever ways to escape and survive. This chapter explores hunting strategies, defense tricks, and the balance that keeps ecosystems thriving.

1622 - Ambush predators hide in cover and strike prey as it passes by.

1623 - Stealth helps many predators catch prey by surprise.

1624 - Cooperative hunters work in teams to corner and capture prey.

1625 - Dolphins coordinate in pods to herd fish into tight schools.

1626 - Orcas use teamwork to trap prey and share the catch within the pod.

1627 - Lions rely on stealth and bursts of speed to take down large herbivores.

1628 - Leopards stalk from cover and spring onto passing animals.

1629 - Tigers hide in tall grass and ambush approaching prey.

1630 - Snow leopards blend into snowy rocks to ambush mountain goats.

1631 - Jaguars deliver a crushing bite to the neck or skull of their prey.

1632 - Pythons coil around prey and squeeze until it stops moving.

1633 - Boas use constriction to subdue their prey as well.

1634 - Crocodiles lie on the riverbank and snap at animals that come to drink.

1635 - Alligators wait patiently with jaws open near water edges to strike.

1636 - Sharks use scent, movement, and electrical senses to locate prey and strike.

1637 - Great white sharks can surge upward from below in a sudden attack.

1638 - Mako sharks are known for fast, agile chases in coastal waters.

1639 - Orcas sometimes create waves to dislodge seals off ice into the water.

1640 - Bears fish the rivers during salmon runs and catch other prey when possible.

1641 - Eagles spot prey from high in the sky using superb vision.

1642 - Hawks dive steeply to grab small mammals from the ground.

1643 - Falcons perform high-speed dives to capture birds on the wing.

1644 - Foxes use keen senses and quick bursts to catch rabbits and mice.

1645 - Wolves communicate during hunts with barks and body language to coordinate moves.

1646 - Hyenas chase and can steal kills from other predators.

1647 - Bears are versatile hunters and opportunistic feeders, adapting to many meals.

1648 - Crows and ravens exploit eggs, nestlings, and carrion with clever tactics.

1649 - Antlions dig funnel traps in sandy soil to snare ground-dwelling insects.

1650 - Praying mantises strike with lightning-fast forelegs to catch prey.

1651 - Spiders spin webs to trap insects and inject venom to paralyze them.

1652 - Scorpions use venomous stings to subdue prey and defend themselves.

1653 - Tarantulas wait patiently to ambush passing insects and small animals.

1654 - Jellyfish drift with currents and use tentacles to catch prey.

1655 - Cuttlefish rapidly change color and pattern to hide or confuse predators and prey.

1656 - Squid snap their tentacles onto prey with quick strikes.

1657 - Sea anemones trap small fish and plankton with stinging tentacles.

1658 - Nudibranchs defend themselves with chemicals and bright warning colors.

1659 - Sea urchins protect themselves with long, sharp spines that deter predators.

1660 - Crabs grab prey with strong claws and drag it away from danger.

1661 - Lobsters use powerful claws to catch prey and defend themselves.

1662 - Snapping shrimp create loud sounds and shockwaves that stun small prey.

1663 - Seahorses hunt tiny crustaceans with elongated snouts.

1664 - Burrowing animals hide from predators by staying underground.

1665 - Sea cucumbers can eject sticky toxins when attacked to deter predators.

1666 - Starfish pry open shells of mollusks and feed on the soft inside.

1667 - Turtles retreat into hard shells to survive predator attacks.

1668 - Armadillos curl into balls, protecting their softer underbelly.

1669 - Porcupines present a formidable quill barrier for many attackers.

1670 - Many insects use cryptic coloring to blend into leaves and bark.

1671 - Some prey species travel in groups to increase alertness and safety.

1672 - Alarm calls warn others and can deter a predator from approaching.

1673 - The "confusion effect" helps prey survive when a group swirls together.

1674 - Seasonal changes shift when predators hunt, influencing prey behavior.

1675 - Camouflage lets prey vanish into grass, bark, or snow.

1676 - Coloration on prey can signal toxicity or unpalatability to predators.

1677 - Playing dead helps some prey avoid further harm after an attack.

1678 - Burrowing animals escape danger by slipping underground.

1679 - Nighttime predators rely on enhanced senses to hunt in the dark.

1680 - Predator-prey interactions shape the evolution of many species.

1681 - Migration can affect how predators find and catch their prey.

1682 - Predator pressure can shape where prey feed or rest.

1683 - Some predators change hunting tactics as prey ages or grows stronger.

1684 - Humans influence predator-prey dynamics by hunting, conserving, or habitat change.

1685 - Electric eels generate shocks to stun prey in muddy river channels.

1686 - Sawfish use long, saw-like snouts to slash at prey.

1687 - Moray eels hide in crevices and ambush prey that pass by.

1688 - Puffers rely on toxins in their skin to deter predators.

1689 - Bioluminescent organisms in deep oceans lure prey with light signals.

1690 - Archerfish and certain reef fish rely on water currents to move prey toward them.

1691 - Juvenile predators learn hunting tricks from watching older hunters.

1692 - Some animals hunt by scent trails to locate hidden prey.

1693 - The smell of blood can attract other predators to a kill.

1694 - Predators often time their strikes for when prey is most vulnerable, such as near water.

1695 - Predators adapt their hunting to different habitats, from forests to oceans.

1696 - Prey species can change their activity patterns to reduce encounters with predators.

1697 - Group living prey can spot danger faster thanks to more eyes and ears.

1698 - The presence of predators influences where prey feed, rest, and travel.

1699 - Predator-prey interactions drive the evolution of new defenses and tricks.

1700 - The dance of hunting and escaping keeps ecosystems dynamic and balanced.

1701 - Some predators use mimicry to resemble harmless or dangerous species to fool prey.

1702 - Prey's vigilance and flight responses improve survival as landscapes change.

1703 - The ability to digest bone and tough tissue helps many predators extract nutrients.

1704 - Some predators specialize in particular prey types to maximize success.

1705 - Prey often combine camouflage, speed, and caution to survive.

1706 - The best hunters study their prey's habits and environments.

1707 - Seasonal migrations of prey can open windows for opportunistic hunters.

1708 - Predators that succeed often use a mix of stealth, speed, and social tactics.

1709 - A successful hunt often involves careful planning and learning from past attempts.

1710 - Predation levels help shape population cycles in nature.

1711 - The wild world thrives on the constant push and pull of hunting, escaping, and surviving.

ODDBALLS AND
UNBELIEVABLE BODIES

Step into a world where animal bodies do things you'd never expect. This chapter dives into extra-long tongues, see-through bodies, detachable tails, strange teeth, and other "how is that real?" adaptations that spark curiosity and wonder.

1712 - A giraffe's tongue can stretch to about 18 inches, letting it pull leaves from the tallest trees.

1713 - An anteater's tongue can extend up to about two feet, sweeping up ants and termites with sticky saliva.

1714 - Pangolins use long, sticky tongues to probe ant and termite nests for tasty morsels.

1715 - Hummingbirds have extendable tongues with forked grooves that help them lap nectar quickly.

1716 - The Komodo dragon samples scents by flicking its forked tongue to bring scent particles to its Jacobson's organ.

1717 - The blue whale has a massive tongue that helps move huge mouthfuls of krill toward the throat during feeding.

1718 - The tail that regrows after shedding is often a different color or texture from the original.

1719 - Starfish can shed an arm to escape a predator and later regrow the arm.

1720 - Sea cucumbers can eject their internal organs as a defense, then regenerate them later.

1721 - Glass catfish are almost perfectly transparent, sometimes revealing their skeletons as they swim.

1722 - The glass octopus is so clear that its organs can be seen through its body in some species.

1723 - Some jellyfish are so transparent that they seem nearly invisible in the water.

1724 - Turritopsis dohrnii, the immortal jellyfish, can revert to a juvenile stage after reaching adulthood.

1725 - The tuatara has a tiny third eye on the top of its head that helps sense light and dark cycles.

1726 - Electric eels can generate powerful electric shocks to stun prey and communicate.

1727 - Electric rays can discharge electricity for defense and hunting.

1728 - Pistol shrimps snap their claws so fast that they create a cavitation bubble and a shockwave.

1729 - Narwhals have long tusks that are actually elongated canine teeth.

1730 - Parrotfish have beak-like teeth formed by fused dental plates that grind coral and algae.

1731 - Lampreys have circular rows of sharp teeth around a suction-cup mouth.

1732 - Sharks continually shed and replace teeth, keeping their bite sharp.

1733 - The hammerhead shark's wide head helps it pin down prey and scan the ocean better.

1734 - The great white shark has multiple rows of teeth that move forward as older ones fall out.

1735 - The rostrum of a sawfish is covered with tiny teeth that help it sense prey in the water.

1736 - The blue dragon sea slug can steal stinging cells from its prey and use them for its own defense.

1737 - Vampire squid uses bioluminescence and a cloak-like web to hide in the deep ocean rather than attacking with teeth.

1738 - The octopus has three hearts and blue blood, which helps it survive in cold, low-oxygen waters.

1739 - The giant squid uses two long feeding tentacles in addition to its other arms to capture prey.

1740 - The firefly squid emits light through photophores to attract prey and mates in the deep sea.

1741 - The lanternfish family glows with bioluminescent organs that help lure prey and avoid danger.

1742 - The hagfish produces slime so thick it can clog a predator's gills, giving the fish a chance to escape.

1743 - The electric ray's body can store electricity in specialized organs called electrocytes for defense.

1744 - The archerfish shoots jets of water to knock insects off plants and into the water to eat them.

1745 - The platypus can swim with its eyes, ears, and nostrils closed, relying on electroreception and touch.

1746 - Some deep-sea fishes have mouths that can stretch open unnaturally wide to swallow large prey.

1747 - The blobfish's unusual shape makes it look "odd" on land but perfectly suited to its deep-sea habitat.

1748 - The hydromedusa jellyfish family uses pulsating movement and light to confuse predators.

1749 - The frill-necked lizard opens a bright, lacy frill to scare away predators.

1750 - The mangrove rivulus can survive out of water for extended periods by absorbing oxygen through its skin.

1751 - The tardigrade, a tiny water-dwelling creature, can survive extreme temperatures, pressure, and radiation.

1752 - The bombardier beetle's spray can produce a visible plume and a warm explosion-like sound.

1753 - The arctic sea butterfly has a translucent body and uses its delicate wings to skim water.

1754 - The rain frog's skin can secrete toxins that deter predators when touched.

1755 - Some sea urchins can regrow damaged spines after injury.

1756 - The mantis shrimp's eye has trinocular vision and can detect a wide spectrum of colors, including polarized light.

1757 - The blue whale's baleen plates are fringed with bristles, enabling efficient filtering of tiny organisms.

1758 - The green sea turtle uses its strong beak to bite through tough seagrass and algae mats.

1759 - The peacock mantis shrimp's colorful body can display a range of wavelengths that communicate with others.

1760 - The pit viper's fangs are folded back against the roof of its mouth and pivot forward when striking.

1761 - Some octopuses can solitarily solve puzzles and escape from enclosures through small openings.

1762 - The naked mole-rat lives underground in nearly zero-light conditions and shows unusual social organization for a rodent.

1763 - The cassowary's helmet-like head features a casque that may help display dominance or sense sound vibrations.

1764 - The lyre snail uses a specialized radula to scrape algae from hard surfaces.

1765 - The sea otter uses its chest as a table to crack shellfish with stones, a behavior known as "stone-roasting."

1766 - The dugong's tusk-like incisors appear mainly in males and are used in mating displays.

1767 - The axolotl's regenerative abilities extend beyond limbs to parts of the brain and spinal cord in some studies.

1768 - The porcupinefish puffs up its body by inflating with water to deter predators.

1769 - Some sharks give birth to live young with functional fins and teeth already developed.

1770 - The African elephant's trunk acts as a hand, nose, and powerful lifting tool in a single structure.

1771 - Seahorses are the only fish in which the male becomes pregnant and gives birth to the young.

1772 - Sea cucumbers can eject long, sticky cuvierian tubules to distract predators, then regenerate them.

1773 - Cuttlefish have eyes with a distinctive W-shaped pupil that helps them judge depth in low light.

1774 - Flying snakes can glide through the air by flattening their bodies and steering with their tails.

1775 - Moray eels have a second set of jaws called pharyngeal jaws to pull prey into their throats.

1776 - Goblin sharks can protrude their jaws forward to grab prey in tight spaces.

1777 - Red-lipped batfish can 'walk' along the sea floor using its modified fins.

1778 - Frilled sharks have long, eel-like bodies with numerous teeth for catching prey in deep water.

1779 - Some sharks and rays have electroreceptors that let them sense the electrical signals of nearby prey.

1780 - Certain turtles can breathe underwater by absorbing oxygen through their cloaca, letting them stay submerged longer.

1781 - Nautilus uses gas-filled chambers inside its shell to control buoyancy and depth.

1782 - Pangolins' scales are made of keratin and can curl into a tight ball for protection.

1783 - Cone snails hunt with a venomous harpoon-like tooth that injects toxins into prey.

1784 - The octopus can taste with its suckers and even detect flavors by touch.

1785 - Crocodiles continually replace teeth throughout life, keeping their bite sharp.

1786 - The star-nosed mole has a star-shaped nose with 22 fleshy tentacles that help it feel its way through the soil and detect prey in the dark.

AFTERWORD

As you close this book, you're not ending a story—you're stepping back into the wild, with questions to chase and marvels to share.

In Incredible Animal Facts for Smart Kids, you've met record-breakers, camouflage masters, and brainy problem-solvers who show that animals aren't passive survivors—they're clever, adaptable teammates.

From super senses to sneaky camouflage, from ocean giants to desert survivors, this book shows how animals adapt, communicate, and survive.

Some moments will stick: the beetle that carries awe-inspiring strength; the cheetah that becomes a blur of speed; the octopus that can change color in a blink; and the glow of bioluminescent deep-sea life.

These surprises come from the big ideas—habitats, senses, and communication—each revealing how life on Earth fits together. You learned to notice tiny clues and big patterns.

Short pages, big ideas, bright pictures: this book is built for curious minds at different speeds, from quick-read fans to those who want to pause and chat about a fact.

Take what you learned on a nature walk, a road trip, or a quiet night at home, and turn it into questions you can explore with friends.

If you loved the journey, please leave a review on the platform you purchased from, and check out the author's other books wherever you buy books. Your thoughts help other curious kids find adventures in science. Keep exploring, stay curious, and let the animal world surprise you again and again.